Food Quality Assurance

Principles and Practices

Food Quality Assurance

Principles and Practices

Inteaz Alli

CRC PRESS

Boca Raton   London   New York   Washington, D.C.

Library of Congress Cataloging-in-Publication Data

Alli, Inteaz.
    Food quality assurance : principles and practices / Inteaz Alli.
        p. cm.
    Includes bibliographical references and index.
    ISBN 1-56676-930-2 (alk. paper)
        1. Food industry and trade—Quality control. I. Title.

TP372.5.A46 2003
664′.068′5—dc21   2003043979

Visit the CRC Press Web site at www.crcpress.com

© 2004 by CRC Press LLC

No claim to original U.S. Government works
International Standard Book Number 1-56676-930-2
Library of Congress Card Number 2003043979
Printed in the United States of America 3 4 5 6 7 8 9 0
Printed on acid-free paper

Dedication

Dedicated to

My wife, my two sons, and all my family

Preface

This book has been prepared with three objectives. First, the book is intended to be used as a textbook for students taking a college- or university-level food quality assurance or food quality management course for the first time. The author's primary objective was to compile information that food science students are expected to be familiar with as part of their college or university program, before they seek career positions in the food industry. Second, it is expected that the book will be used by food plant employees who have not had prior training in food quality and food safety. Third, the book will be useful to food industry quality practitioners who need to become familiar with updated information relating to their work, including the evolution of principles, practices, and terminology in food quality assurance and food quality management.

There have been substantial changes in food industry quality and safety practices since the beginning of the 1990s. These changes have been driven primarily by the adoption of the *Hazard Analysis Critical Control Point (HACCP)* system for ensuring food safety worldwide, as well as the widespread use of the International Organization for Standardization's quality standard *(ISO 9001:2000)* by various industry sectors, including the food industry. Consequently, it has become essential for students preparing for careers in the food industry to learn about these systems as part of the curricula in their university or college programs.

The contents of this textbook are devoted primarily to food quality systems and food safety systems. The goal is to introduce the principles, practices, and vocabulary of food quality and food safety in a manner that is understandable to a college or university food science and technology student, as well as to a novice to the area of food quality and food safety.

The primary sources of information for the book are recent versions of documents available to the public from recognized international and national organizations. These include the Codex Alimentarius Commission, the ISO, the American Society for Quality, and the National Advisory Committee for Microbiological Criteria for Foods. In addition, information contained in documents of various government regulatory agencies responsible for food quality and food safety in North America have been used.

The book presents the principles and practices of food industry good manufacturing practices (GMPs), HACCP, and the ISO 9000 quality management system standards.

Acknowledgments

The author expresses sincere thanks to the following:

- Alam Alli, Razan El-Ramahi, and Terri Cundi — students who read chapter drafts and provided comments from a student perspective
- Paul Brennan, Mike Hudson, Laurent Laflamme, and Renee Sauvageau — food industry quality assurance professionals who reviewed chapter drafts for technical content relating to food quality and food safety — and Sam Weissfelner — quality management systems expert for reviewing Chapters 1 and 3 at the proof stage.
- Barry Callebaut North America Inc., International Stockfood Corp., Lallemand Inc., and SGS/ICS Canada Inc. — companies that provided the author the opportunity to practice in the field of quality assurance and quality management.
- ASQ Quality Press for permission to use definitions in Chapter 1 from the following sources:
    - "Quality Glossary," *Quality Progress*, July 2002, 49–61.
    - ASQ Food, Drug and Cosmetic Division, *Food Processing Industry Quality Systems Guidelines*, ASQ Quality Press, Milwaukee.
- The Food and Agricultural Organization of the United Nations for permission to use materials from the Recommended International Code of Practices, General Principles of Food Hygiene, Codex Alimentarius Commission, Joint FAO/WHO Food Standards Program, 1997.
- *Journal of Food Protection*. Copyright held by the International Association for Food Protection, Des Moines, Iowa, U.S., for permission to use material from NACMCF.
- IHS Canada/Standards Council of Canada, Ottawa, Ontario, Canada, for permission to use materials from the ISO 9000 Standards.

About the author

Inteaz Alli, Ph.D., is a Professor of Principles of Food Quality Assurance and Food Analysis at the Department of Food Science and Agricultural Chemistry, McGill University, Montreal, Quebec, Canada. He served as Chair of the Department from 1995 to 2003. He is a Lead Auditor (National Quality Institute) for Quality Management Systems, a Certified Quality Auditor and Hazard Analysis Critical Control Point Auditor (American Society for Quality), and a Professional Chemist. Dr. Alli has served in numerous executive positions in the Montreal Section of the American Society for Quality, and in several editorial and committee positions in the Canadian Institute of Food Science and Technology and the Quality Assurance Division of the Institute of Food Technologists. The author has also served as an advisor and consultant for both public- and private-sector organizations in the food industry.

Contents

chapter one

Vocabulary of food quality assurance

1.1 Introduction

This chapter introduces the vocabulary of food quality assurance. The terms and phrases defined and explained here are the common vocabulary of food quality and food safety. Knowing and understanding the vocabulary and technical language of a field is essential for any area of study.

The international recognition of systems for food safety and for quality management has resulted in the need to adopt terminology that can be interpreted in a uniform and consistent manner. The systems for food safety and for quality management that have been adopted by international organizations such as the Codex Alimentarius Commission and the International Organization for Standardization (ISO) have been based on fundamental principles developed by recognized experts or recognized scientific or professional organizations. As a result, there is now standardized vocabulary in the field of food quality assurance.

When the ISO revised the international quality management system standards in 2000, a section of the ISO 9000:2000 standard was devoted to vocabulary. The definitions in the ISO 9000:2000 standard are the most current and widely accepted definitions. In the area of food safety, the Codex Alimentarius Commission has defined terms used in the internationally recognized hazard analysis critical control point (HACCP) system. These definitions are the most widely accepted internationally and are similar to — and in many cases, identical to — the definitions of the National Advisory Committee on Microbiological Criteria for Foods (NACMCF).

The role of the various organizations in promoting the use of a consistent vocabulary in systems for food quality and food safety is recognized in this chapter. In general, the definitions and explanations given are those of the ISO, the American Society for Quality (ASQ), the Codex Alimentarius Commission and NACMCF. The definitions and explanations of terms and phrases as they are stated by these four organizations are provided here in alphabetical order, with permission.

1.2 *Definitions and explanation of terms**

Acceptable quality level (AQL) In a continuing series of lots, a quality level that, for the purpose of sampling inspection, is the limit of satisfactory process average (QP, 2002).

    The AQL is used in acceptance sampling of product lots. It can also be expressed as the maximum percent defective that can be considered satisfactory for the average of the process that is producing several lots of a product.

Acceptance sampling Procedure by which a decision to accept or reject a lot is based on the results of inspection of samples (ASQ, 1998).

American Society for Quality (ASQ) A professional, not-for-profit association that develops, promotes, and applies quality-related information and technology for the private sector, government, and academia (QP, 2002).

    ASQ is the largest professional association for practitioners in the quality field. Its activities include offering the following certification programs to quality professionals: Certified Mechanical Engineer, Certified Quality Auditor (CQA), Certified Quality Auditor (CQA)-Biomedical, Certified Quality Auditor (CQA)-Hazard Analysis Critical Control Point (HACCP), Certified Quality Engineer (CQE), Certified Quality Improvement Associate (CQIA), Certified Quality Manager (CQM), Certified Quality Technician (CQT), Certified Reliability Engineer (CRE), Certified Six Sigma Black Belt (CSSBB), Certified Software Quality Engineer (CSQE).

Assignable cause Those observations of a process measurement that are not randomly caused (ASQ, 1998). A name for the source of variation in a process that is not due to chance and, therefore, can be identified and eliminated. Also called *special cause* (QP, 2002).

    Assignable causes can be identified by use of control charts.

Attributes (method of) A measurement of quality consisting of noting the presence (or absence) of some characteristic or attribute in each of the units in a group under consideration, and counting how many units do (or do not) possess the quality attribute or how many such events occur in the unit, group, or area (ASQ, 1998). See also: *variables*.

---

* Source: "Quality Glossary." ASQ *Quality Progress*, July 2002, Pages 49-61. Reprinted with permission. Source: ASQ Food, Drug, and Cosmetic Division. 1998. *Food Processing Industry Quality System Guidelines*. Milwaukee, WI: ASQ Quality Press. Reprinted with permission. From the Food and Agriculture Organization of the United Nations, Codex Alimentarius, 1997 *Basic Texts on Food Hygiene*, Codex Alimentarius Commission, Joint FAO/WHO Food Standards Program, Rome. From ISO 9000: 2000 *International Standard*, 2nd ed., 2002-12-15, *Quality management systems — fundamentals and vocabulary*. ISO, Geneva. From *J. Food Prot.*, 61, 9, 1998, 1246-1259.

Audit Systematic, independent, and documented process for obtaining audit evidence and evaluating it objectively to determine the extent to which audit criteria are fulfilled (ISO 9000:2000).

Common types of audits in the food industry are audits of quality management systems, audits of good manufacturing practices (GMP audits), and audits of the hazard analysis critical control point system (HACCP audit). GMP audits and HACCP audits are food safety audits and are frequently conducted in food companies by government regulatory agencies. Internal audits are commonly carried out within a food company for the benefit of the company. Internal audits are first-party audits. External audits are also common in the food industry. An external audit conducted in a food company by a customer, or on behalf of the customer, for the benefit of the customer is a second-party audit. An external audit carried out by an organization that is independent of both the company and the customer is a third-party audit. A registration audit is a third-party audit carried out for the purpose of registering the company to a recognized standard, such as the ISO 9001:2000 quality management system standard.

Auditee Organization being audited (ISO 9000:2000).

For an internal audit, the auditee is commonly a department, section, or sector that is being audited. For an external audit, the auditee is the company that is being audited.

Auditor A person with the competence to conduct an audit (ISO 9000:2000). For an internal audit, the auditor is appointed by the company's management. For an external audit, the auditor is appointed by the organization responsible for conducting the audit. For all types of audits, an auditor must have the necessary qualifications to conduct the audit. The person who performs an audit of a process must not be involved in the activities of the process; this ensures that the audit is an independent process.

Audit client Organization or person requesting an audit (ISO 9000:2000). For example, in the case of a second-party audit of an organization, done by someone on behalf of a customer, the audit client is the customer.

Audit conclusion Outcome of an audit provided by an audit team after consideration of the audit objectives and all audit findings (ISO 9000:2000).

The audit conclusions are contained in the audit report after the completion of an audit. In general, when there is more than one person on the audit team, the lead auditor on the audit team is responsible for providing the audit conclusions.

Audit criteria Set of policies, procedures, or requirements used as a reference (ISO 9000:2000).

Audit evidence Records, statements of fact, or other verifiable information relevant to the audit criteria (ISO 9000:2000).

Audit evidence is obtained by an auditor during an audit by examination of records, from the observation of activities, and from information obtained during interviews of personnel responsible for activities.

Audit findings Results of the evaluation of the collected audit evidence against audit criteria (ISO 9000:2000).

After gathering the audit evidence during the course of an audit, an auditor carries out an objective evaluation of the evidence in relation to the audit criteria to arrive at the audit findings.

Audit program Set of one or more audits planned for a specific time frame and directed toward a specific purpose (ISO 9000:2000).

Audit team One or more auditors conducting an audit (ISO 9000:2000).

An audit team should have a lead auditor who has overall responsibility for the audit.

Batch See: *lot*.

Benchmarking An improvement process in which a company measures its performance against that of best-in-class companies, determines how those companies achieved their performance levels, and uses the information to improve its own performance. The subjects that can be benchmarked include strategies, operations, processes, and procedures (QP, 2002).

Best practice A superior method or innovative practice that contributes to the improved performance of an organization, usually recognized as "best" by other peer organizations (QP, 2002).

Blemish An imperfection severe enough to be noticed but that should not cause any real impairment with respect to intended normal or reasonably foreseeable use (QP, 2002). See also: *defect, imperfection,* and *non-conformity.*

Body of knowledge (BOK) The prescribed aggregation of knowledge in a particular area an individual is expected to have mastered to be considered or be certified as a practitioner (QP, 2002).

Breakthrough improvement A dynamic, decisive movement to a new, higher level of performance (QP, 2002).

Calibration Correcting a measuring device to agree with a standard (ASQ, 1998). The comparison of a measurement instrument or system of unverified accuracy to a measurement instrument or system of a known accuracy to detect any variation from the required performance specification (QP, 2002).

Capability Ability of an organization, system, or process to realize a product that will fulfil the requirements for that product (ISO 9000:2000). See also: *process capability.*

Cause-and-effect diagram Used for analyzing process dispersion, it illustrates the main causes and subcauses leading to an effect (symptom). It was developed by Kaoru Ishikawa and is also referred to as the *Ishikawa diagram* and the *fishbone diagram* because of its shape. It is one of the seven tools of quality (QP, 2002).

Characteristic Distinguishing feature (ISO 9000:2000). A property that helps to differentiate among items of a given sample or population (ASQ, 1998). The factors, elements, or measures that define and differentiate a process, function, product, service, or other entity (QP, 2002). See also: *quality characteristic.*

For food products, some common quality characteristics are safety, suitability, appearance, color, taste, texture, and composition.

Chart A tool for organizing, summarizing and depicting data in graphic form (QP, 2002).

Check sheet A form for recording data from a process or product; may be used as a first step of a processor product quality analysis (ASQ, 1998). It is a simple data recording device custom designed by the user to allow ready interpretation of the data recorded. It is one of the seven tools of quality (QP, 2002).

Checklist A tool used to ensure that all important steps or actions in an operation have been taken. A checklist contains items that are important or relevant to an issue or situation (QP, 2002).

Checklists are commonly used by auditors as a tool to obtain objective evidence during audits.

Cleaning The removal of soil, food residue, dirt, grease, or other objectionable matter (Codex Alimentarius, 1997).

Codex Alimentarius Set of internationally recognized food standards developed by the Codex Alimentarius Commission.

Codex Alimentarius Commission A joint, subsidiary body of the Food and Agricultural Organization (FAO) of the United Nations and the World Health Organization (WHO). The Codex Alimentarius Commission was established in 1962 for the purpose of developing, promoting, and harmonizing standards for food in order to facilitate international trade. Membership on the Commission is open to all member nations and associate members of the FAO and

WHO. The Codex Alimentarius Commission has developed numerous internationally recognized food standards as well as an internationally recognized HACCP system.

Common cause(s) Attributes or variables of a process that are the result of random variations inherent in the process (ASQ, 1998). Causes of variation that are inherent in a process over time. They affect every outcome of the process and everyone working in the process (QP, 2002).

Common causes are also referred to as chance causes. See also: *assignable cause* and *special causes*.

Company culture A system of values, beliefs and behaviors inherent in a company. To optimize business performance, top management must define and create the necessary culture (QP, 2002).

Competence Demonstrated ability to apply knowledge and skills (ISO 9000:2000).

Compliance The state of an organization that meets prescribed specifications, contract terms, regulations, or standards (QP, 2002). See also: *conformance* and *conformity*.

Concession Permission to use or release a product that does not conform to specified requirements (ISO 9000:2000). See also: *deviation permit*. A customer can grant a concession to an organization for the acceptance of a product that has been made but which does not conform to the product requirements; the concession is granted before use or release of the nonconforming product.

Conformance An affirmative indication or judgement that a product has met the requirements of a relevant specification, contract, or regulation (QP, 2002). See also: *compliance* and *conformity*.

The terms *conformity* and *conformance* are often used interchangeably. In the ISO 9000:2000 standards, *conformity* is used in preference to *conformance*.

Conformity Fulfillment of an requirement (ISO 9000:2000). See also *compliance* and *conformance*.

Conformity can be associated with a product, a process, or a system.

Consumer The external customer to whom a product or service is ultimately delivered. Also called end user (QP, 2002). See also: *customer*.

Consumer's risk The probability that an unacceptable lot will be accepted under a specified sampling plan (ASQ, 1998). Pertains to sampling and the potential risk that bad product will be accepted and shipped to the consumer (QP, 2002).

Contaminant Any biological or chemical agent, foreign matter, or other substances not intentionally added to food, which may compromise food safety or food suitability (Codex Alimentarius, 1997).

Contamination The introduction or occurrence of a contaminant in food or food environment (Codex Alimentarius, 1997). See also: *cross-contamination*.

Continual improvement Recurring activity to increase the ability to fulfill requirements (ISO 9000:2000). Continual improvement is one of the eight quality management principles recognized by ISO 9000:2000. This principle states: "Continual improvement of the organization's overall performance should be a permanent objective of the organization."

Continuous improvement (CI) Sometimes called continual improvement. The ongoing improvement of products, services, or processes through incremental and breakthrough improvements (QP, 2002). See also: *continual improvement*.

Control (Verb) To take all necessary actions to ensure and maintain compliance with the criteria established in the HACCP plan. (Noun) The state wherein correct procedures are being followed and criteria are being met (Codex Alimentarius, 1997). To manage the conditions of an operation to maintain compliance with established criteria. The state in which correct procedures are being followed and criteria are being met (NACMCF, 1997). Commonly used in other phrases, including *process control, quality control, control chart, control point, critical control point*, and *control measure*.

Control chart Usually a graph of groups or individuals of a process parameter sampled at regular intervals and plotted on a format that includes statistical process limits (ASQ, 1998). A chart with upper and lower control limits on which values of some statistical measure for a series of samples or subgroups are plotted. The chart frequently shows a central line to help detect a trend of plotted values toward either control limit. It is one of the seven tools of quality (QP, 2002). Control charts are commonly used in statistical process control. There are several types of control charts; the most commonly used are the control chart for variables and the control chart for attributes.

Control limits Statistically calculated extreme values of a process or product, outside of which a process is considered to be out of statistical control; not to be confused with specification limits, tolerance limits, or critical limits (ASQ, 1998). The natural boundaries of a process within specified confidence levels, expressed as the upper control limit (UCL) and the lower control limit (LCL) (QP, 2002). See also: *control chart, upper control limit* and *lower control limit*.

Control measure Any action and activity that can be used to prevent or eliminate a food safety hazard or reduce it to an acceptable level (Codex Alimentarius, 1997). Any action or activity that can be used to prevent, eliminate or reduce a significant hazard (NACMCF, 1997).

Control measures are identified to control food safety hazards during the development of a HACCP plan.

Control point Any step at which biological, chemical or physical factors can be controlled (NACMCF, 1997; ASQ, 1998). See also: *critical control point*.

Correction Action to eliminate a detected nonconformity (ISO 9000:2000). See also: *corrective action*.

For a product with a detected nonconformity, the correction can be to rework the product so that it conforms to the requirements.

Corrective action Action to eliminate the cause of a detected nonconformity or other undesirable situation (ISO 9000:2000). Any action to be taken when the results of monitoring at the CCP indicate a loss of control (Codex Alimentarius, 1997). Procedures followed when a deviation occurs (NACMCF, 1997). The implementation of solutions resulting in the reduction or elimination of an identified problem (QP, 2002). See also: *correction* and *preventive action*.

Critical control point (CCP) A step at which control can be applied and is essential to prevent or eliminate a food safety hazard or reduce it to an acceptable level (Codex Alimentarius, 1997; NACMCF, 1997). The processing factors whose loss of control would result in an unacceptable food safety risk (ASQ, 1998). See also: *control point*.

Critical control point (CCP) decision tree A sequence of questions to assist in determining whether a control point is a CCP (NACMCF, 1997). CCP decision trees have been developed by the NACMCF and by Codex Alimentarius.

Critical limit A criterion that separates acceptability from unacceptability (Codex Alimentarius, 1997). A maximum or minimum value to which a biological, chemical, or physical parameter must be controlled at a CCP to prevent, eliminate, or reduce to an acceptable level, the occurrence of a food safety hazard (NACMCF, 1997).

Criterion A requirement on which a judgment or decision can be made (NACMCF, 1997).

Critical processes Processes that present serious potential dangers to human life, health, and the environment or that risk the loss of very large sums of money or a great number of customers (QP, 2002).

Cross-contamination The transfer of contaminant from one food to another; common means of cross-contamination include personnel, utensils and equipment.

Customer Organization or person that receives a product (ISO 9000:2000). See also: *internal customer and external customer.*

A customer can be a consumer, a client, an end-user, a retailer, a beneficiary, or a purchaser. A customer can be internal or external to an organization.

Customer focus One of the eight quality management principles recognized by ISO 9000:2000, states: "Organizations depend on their customers and, therefore, should understand current and future customer needs, should meet customer requirements and strive to exceed customer expectations."

Customer satisfaction Customer's perception of the degree to which the customer's requirements have been fulfilled (ISO 9000:2000).

Data A set of collected facts. There are two basic kinds of numerical data: measured or variable data and counted or attribute data (QP, 2002).

Defect Nonfulfillment of a requirement related to an intended or specified use (ISO 9000:2000). A departure of a quality characteristic from its intended level or state that occurs with a severity sufficient to an associated product or service not to satisfy intended normal, or reasonably foreseeable, usage requirements (ASQ, 1998). A product's or service's nonfulfillment of an intended requirement or reasonable expectation of use, including safety considerations. There are four classes of defects: class one, very serious, leads directly to severe injury or catastrophic economic loss; class two, serious, leads directly to significant injury or significant economic loss; class three, major, is related to major problems with respect to intended normal or reasonably foreseeable use; class four, minor, is related to minor problems with respect to intended normal or reasonably foreseeable use (QP, 1992). See also: *nonconformity.*

Defective A unit of product containing at least one defect or having several imperfections that in combination cause the unit not to satisfy normal or reasonably foreseeable usage requirements (ASQ, 1998). A defective unit; a unit of product that contains one or more defects with respect to the quality characteristic(s) under characterization (QP, 2002).

Deming cycle See: *plan-do-check-act cycle.*

Deming's 14 points W. Edwards Deming's 14 management practices to help companies increase their quality and productivity:
1. Create constancy of purpose for improving products and services.

2.   Adopt the new philosophy.

3.   Cease dependence on inspection to achieve quality.

4.   End the practice of awarding business on price alone; instead, minimize total cost by working with a single supplier.

5.   Improve constantly and forever every process for planning, production and service.

6.   Institute training on the job.

7.   Adopt and institute leadership.

8.   Drive out fear.

9.   Break down barriers between staff areas.

10. Eliminate slogans, exhortations and targets for the workforce.

11. Eliminate numerical quotas for the workforce and numerical goals for management.

12. Remove barriers that rob people of pride of workmanship, and eliminate the annual rating or merit system.

13. Institute a vigorous program of education and self-improvement for everyone.

14. Put everybody in the company to work to accomplish the transformation (QP, 2002).

Dependability Collective term used to describe the availability performance and its influencing factors: reliability performance, maintainability performance and maintenance support performance (ISO 9000:2000).

Design and development Set of processes that transforms requirements into specified characteristics or into the specification of a product, process or system (ISO 9000:2002).

Deviation Failure to meet a critical limit (Codex Alimentarius, 1997; NAC-MCF, 1997). See also *nonconformity.*

The term deviation is used in the HACCP vocabulary; it represents a situation where there is a failure to meet a critical limit that has been established as a requirement at a process step that is a CCP.

Deviation permit Permission to depart from the originally specified requirements of a product prior to realization (ISO 9000:2000). See also *concession.*

A deviation permit from a customer allows an organization to realize a product without having to meet the customer's specified requirements. Generally, a deviation permit applies to a specified quantity of product or a specified period of time, and is for a specific use of the product.

Disinfection The reduction, by means of chemical agents and/or physical methods, of the number of microorganisms in the environment, to a level that does not compromise food safety or food suitability (Codex Alimentarius, 1997).

Document Information and its supporting medium (ISO 9000:2000). See also *record*.

> The types of documents that are part of a quality management system include specifications, records, documented procedures, work instructions, drawings, reports, standards, quality manuals, and quality plans. The supporting medium for documents includes paper, photograph, and magnetic, electronic or optical computer disc.

Effectiveness Extent to which planned activities are realized and planned results achieved (ISO 9000:2002). The state of having produced a decided upon or desired effect (QP, 2002). See also: *efficiency.*

Efficiency Relationship between the result achieved and the resources used (ISO 9000:2002). The ratio of the output to the total input in a process (QP, 2002). See also: *effectiveness.*

Employee involvement A practice within an organization whereby employees regularly participate in making decisions on how their work areas operate, including making suggestions for improvement, planning, goal setting, and monitoring performance (QP, 2002). See also: *involvement of people.*

Empowerment A condition whereby employees have the authority to make decisions and take action in their work areas without prior approval (QP, 2002).

Establishment Any building or area in which food is handled and the surroundings are under the control of the same management (Codex Alimentarius, 1997).

Expectations (customer) Customer perceptions about how an organization's products and services will meet the customer's specific needs and requirements (QP, 2002).

External customer A person or organization that receives a product, service or information but is not part of the organization supplying the product (QP, 2002). See also: *customer* and *internal customer.*

External failure Nonconformance identified by the external customers (QP, 2002). See also: *internal failure.*

Factual approach to decision making This is one of the eight quality management principles recognized by ISO 9000:2000. This principle states: "Effective decisions are based on the analysis of data and information."

Fishbone diagram See: *cause-and-effect diagram.*

Flowchart A graphical representation of the steps in a process, which is drawn in order to better understand the process. It is one of the seven tools of quality (QP, 2002). See also: *flow diagram*.

Flow diagram A systematic representation of the sequence of steps or operations used in the production or manufacture of a particular food item (Codex Alimentarius, 1997). A pictorial representation of a process indicating each of the branches and main steps in order of performance (ASQ, 1998). See also: *flowchart*.

The preparation of a process flow diagram is one of the steps in the development of an HACCP plan for a food item.

Food handler Any person who directly handles packaged or unpackaged food, food equipment and utensils, or food contact surfaces and is therefore expected to comply with food hygiene requirements (Codex Alimentarius, 1997).

Food hygiene All conditions and measures necessary to ensure the safety and suitability of food at all stages of the food chain (Codex Alimentarius, 1997).

Food safety The assurance that food will not cause harm to the consumer when it is prepared and/or eaten according to its intended use (Codex Alimentarius, 1997).

Food suitability Assurance that food is acceptable for human consumption according to its intended use (Codex Alimentarius, 1997).

Go/no-go State of a unit or product. Two parameters are possible: go (conforms to specifications) and no-go (does not conform to specifications) (QP, 2002).

Grade Category or rank given to different quality requirements for products, processes, or systems having the same functional use (ISO 9000:2000).

Hazard A biological, chemical, or physical agent in, or condition of, food with the potential to cause an adverse health effect (Codex Alimentarius, 1997). A biological, chemical, or physical agent that is reasonably likely to cause illness or injury in the absence of its control (NACMCF, 1997).

Hazard analysis The process of collecting and evaluating information on hazards and conditions leading to their presence to decide which are significant for food safety and, therefore, should be addressed in the HACCP plan (Codex Alimentarius, 1997). The process of collecting and evaluating information on hazards associated with the food under consideration to decide which are significant and must be addressed in the HACCP plan (NACMCF, 1997).

Hazard analysis critical control point (HACCP) A system that identifies, evaluates, and controls hazards that are significant for food safety (Codex Alimentarius, 1997). A systematic approach to the identification, evaluation, and control of food safety hazards (NACMCF, 1997).

HACCP, which is recognized for its science-based approach, consists of a set of seven principles that have been adopted internationally through the work of the Codex Alimentarius Commission.

HACCP plan A document prepared in accordance with the principles of HACCP to ensure control of hazards that are significant for food safety in the segment of the food chain under consideration (Codex Alimentarius, 1977). The written document that is based on the principles of HACCP and that delineates the procedures to be followed (NACMCF, 1997).

An HACCP plan is obtained after an HACCP team has carried out the activities required by the seven principles of HACCP.

HACCP system The result of the implementation of the HACCP plan (NACMCF, 1997).

HACCP team The group of people responsible for developing, implementing, and maintaining the HACCP system (NACMCF, 1997).

Histogram A bar chart representing a frequency distribution of a process variable or product characteristic (ASQ, 1998). A graphic summary of variation in a set of data. It is one of seven tools of quality (QP, 2002).

Imperfection A quality characteristic's departure from its intended level or state without any association to conformance to specification requirements on usability of a product or service (QP, 2002). See also: *blemish, nonconformity* and *defect*.

Improvement The positive effect of a process change effort (QP, 2002). See also: *breakthrough improvement* and *incremental improvement*.

In-control process A process in which the statistical measure being evaluated is in a state of statistical control; in other words, the variations among the observed sampling results can be attributed to a constant system of chance causes (QP, 2002). See also: *out-of-control process*.

Incremental improvement Improvements that are implemented on a continual basis (QP, 2002). See also: *breakthrough improvement*.

Information Meaningful data (ISO 9000:2000). See also: *data*.

Infrastructure System of facilities, equipment, and services needed for the operation of an organization (ISO 9000:2000).

Inputs The products, services, materials, and so forth obtained from suppliers and used to produce the outputs delivered to customers (QP, 2002). See also: *outputs.*

Inspection Conformity evaluation by observation and judgment accompanied as appropriate by measurement, testing, or gauging (ISO 9000:2000). The process of measuring, examining, testing, gauging, or otherwise comparing the unit with the applicable requirements (ASQ, 1998). Measuring, examining, testing and gauging one or more characteristics of a product or service and comparing the results with specified requirements to determine whether conformity is achieved for each characteristic (QP, 2002).

Inspection (100%) Inspection of all the units in a lot or batch (QP, 2002).

Inspection lot A collection of similar units or a specific quantity of similar material offered for inspection and acceptance at one time (QP, 2002).

Interested party Person or a group having an interest in the performance or success of an organization (ISO 9000:2000).

The interested parties of an organization include the organization's customers, suppliers, owners, partners, bankers, people in the organization, or anyone who may have an interest in some aspect of the organization's performance.

Internal customer The recipient (person or department) within an organization of an output from another person or department in the organization (QP, 2002). See also: *external customer.*

Internal failure A product failure that occurs before the product is delivered to external customers (QP, 2002). See also: *external failure.*

International Organization for Standardization (ISO) A network of national standards institutes from 140 countries working in partnership with international organizations, governments, industry, business, and consumer representatives to develop and publish international standards (QP, 2002). A worldwide federation of national standard bodies which are referred to as ISO member bodies (ISO 9000:2000).

Involvement of people This is one of the eight quality management principles recognized by ISO 9000:2000. This principle states: "People at all levels are the essence of an organization and their full involvement enables their abilities to be used for the organization's benefit."

ISO 9000 family of standards A coherent set of four quality management system standards facilitating mutual understanding in national and international trade (ISO 9000:2000).

The standards have been developed by the International Organization for Standardization to assist all types and sizes of organizations to implement and operate effective quality management systems. ISO 9000:2000 Quality management systems — Fundamentals and vocabulary describes fundamentals of quality management systems and specifies and defines the terminology used in these systems. ISO 9001:2000 Quality management systems — Requirements specify the requirements for the quality management system of an organization that needs to demonstrate its ability to provide products that meet customer and regulatory requirements, and aims to enhance customer satisfaction. ISO 9004:2000 Quality management systems — Guidelines for performance improvements provides guidelines that address the effectiveness and efficiency of quality management systems to improve organizational performance and the satisfaction of customers and other parties. ISO 19011:2002 provides guidance on auditing quality management systems and environmental management systems.

Just-in-time (JIT) manufacturing An optimal material requirement planning system for a manufacturing process in which there is little or no manufacturing material inventory on hand at the manufacturing site and little or no incoming inspection (QP, 2002).

Just-in-time training The provision of training only when it is needed to all but eliminate the loss of knowledge and skill caused by a lag between training and use (QP, 2002).

Kaizen A Japanese term that means gradual, unending improvement by doing little things better and setting and achieving increasingly higher standards (QP, 2002).

Leadership This is one of the eight quality management principles recognized by ISO 9000:2000. This principle states: "Leaders establish unity of purpose and direction of the organization. They should create and maintain the internal environment in which people can become fully involved in achieving the organization's objectives." An essential part of a quality improvement effort. Organization leaders must establish a vision, communicate that vision to those in the organization, and provide the tools and knowledge necessary to accomplish the vision (QP, 2002).

Lot A definite quantity of product or material accumulated under conditions that are considered uniform for sampling purposes (ASQ, 1998). A defined quantity of product accumulated under conditions considered uniform for sampling purposes (QP, 2002).

Lot (batch) A definite quantity of some product manufactured under conditions of production that are considered uniform (QP, 2002).

Lower control limit (LCL) Control limit for points below the central line in a control chart (QP, 2002). See also: *upper control limit.*

Malcolm Baldrige National Quality Award (MBNQA) An award established by the U.S. Congress in 1987 to raise awareness of quality management and recognize U.S. companies that have implemented successful quality management systems. The awards program is managed by the National Institute of Standards and Technology (NIST) and administered by the American Society for Quality (QP, 2002).

The Malcolm Baldrige National Quality Award is considered an excellence model and the criteria are referred to as the Baldrige Criteria for Performance Excellence; the seven categories of criteria are: leadership, strategic planning, customer and market focus, information and analysis, human resource focus, process management, and results. See also: *quality management principles.*

Management Coordinated activities to direct and control an organization (ISO 9000:2000). See also: *top management* and *quality management.*

Management review A periodic meeting of management at which it reviews the status and effectiveness of the organization's quality management system (QP, 2002).

Management system System to establish policy and objectives and to achieve those objectives (ISO 9000:2000). See also "quality management system."

There are several types of management systems; these include quality management systems, environmental management systems, financial management systems, and occupational health and safety systems. Quality management systems are covered by the ISO 9000:2000 standards.

Measurement The act or process of quantitatively comparing results with requirements (QP, 2002).

Measurement process Set of operations to determine the value of a quantity (ISO 9000:2000). See also: *inspection* and *test.*

Measuring equipment Measuring instrument, software, measurement standard, reference material, or auxiliary apparatus or combination thereof necessary to realize a measurement process (ISO 9000:2000).

Monitor The act of conducting a planned sequence of observations or measurements of control parameters to assess whether a CCP is under control (Codex Alimentarius, 1997). To conduct a planned sequence of observations or measurements to assess whether a CCP is under control and to produce an accurate record for future use in verification (NACMCF, 1997).

Mutually beneficial supplier relationships This is one of the eight quality management principles recognized by ISO 9000:2000. This principle states: "An organization and its suppliers are interdependent and a mutually beneficial relationship enhances the ability of both to create value."

National Institute of Standards and Technology (NIST) An agency of the U.S. Department of Commerce that develops and promotes measurements, standards, and technology. NIST manages the Malcolm Baldrige National Quality Award (QP, 2002).

Nonconformity Nonfulfillment of a requirement (ISO 9000:2000). A departure of a quality characteristic from its intended level or state that occurs with severity sufficient to cause associated product or service to meet a specification requirement (ASQ, 1998). The nonfulfillment of a specified requirement (QP, 2002). See also: *blemish, defect,* and *imperfection*.

Objective evidence Data supporting the existence or verity of something (ISO 9000:2000).

Operating characteristic (OC) curve A plot showing, for a given sample plan, the probability of accepting a lot, as a function of the lot quality (ASQ, 1998).

Organization Group of people and facilities with an arrangement of responsibilities, authorities, and relationships (ISO 9000:2000).

Organizational structure Arrangement of responsibilities, authorities, and relationships between people (ISO 9000:2000).

Out-of-control process A process in which the statistical measure being evaluated is not in a state of statistical control. In other words, the variations among the observed sampling results cannot be attributed to a constant system of chance causes but to assignable or special causes (QP, 2002). See also: *in-process control*.

Outputs Products, materials, services, or information provided to customers from a process (QP, 2002).

Pareto chart A bar chart or graph of the frequency of product or process measurements. The bars are plotted in order of frequency. The graph may be expressed in numbers or cumulative percentages (ASQ, 1998). A graphical tool for ranking causes from most significant to least significant. It is based on the Pareto principle which was first defined by J.M. Juran in 1950. The principle, named after nineteenth-century economist Vilfredo Pareto, suggests that most effects come from relatively few causes; that is, 80% of the effects come from 20% of the possible causes. It is one of the seven tools of quality (QP, 2002).

Pest control Suppression of pests, usually by chemical, physical or environmental means, to a level at which few are still existing (ASQ, 1998).

Plan-do-check-act (PDCA) cycle A four-step process for quality improvement, which is sometimes referred to as the Shewhart cycle and sometimes as the Deming cycle. In the first step (plan), a plan to effect improvement is developed. In the second step (do), the plan is carried out. In the third step (check), the effects of the plan are observed. In the last step (act), the results are studied to determine what was learned and can be predicted. Also called the plan-do-study-act (PDSA) cycle (QP, 1992).

The ISO 9001:2000 standard proposes that the methodology of the PDCA cycle can be applied to all processes in an organization's quality management system, as follows: plan – establish the objectives and processes necessary to deliver results in accordance with customer requirements and the organization's policies; do – implement the processes; check – monitor and measure processes and product against policies, objectives, and requirements for the product and report the results; act – take actions to continually improve process performance.

Prerequisite programs Procedures, including good manufacturing practices that address operational conditions providing the foundation for the HACCP system (NACMCF, 1997).

Prerequisite programs must be developed and implemented before an effective HACCP system can be implemented. The NACMCF provides the following examples of common prerequisite programs: facilities, supplier control, specifications, production equipment, cleaning and sanitation, personal hygiene, training, chemical control, receiving, storage and shipping, traceability and recall, pest control. These topics are covered in the Codex General Principles of Food Hygiene under the following categories: design and facilities, control of operation, maintenance and sanitation, personal hygiene, transportation, product information and consumer awareness, and training.

Preventive action Action taken to eliminate the cause of a potential nonconformity or other undesirable potential situation (ISO 9000:2000). See also: *corrective action*.

Procedure Specified way to carry out an activity or a process (ISO 9000:2000). The steps in a process and how these steps are to be performed for the process to fulfill customers' requirements (QP, 2002).

Process Set of interrelated or interacting activities which transforms inputs into outputs (ISO 9000:2000). A set of interrelated work activities characterized by a set of specific inputs and value-added tasks that make up a procedure for a set of specific outputs (QP, 2002).

Process approach This is one of the eight quality management principles recognized by ISO 9000:2000. This principle states: "A desired result is achieved more efficiently when activities and related resources are managed as a process." The systematic identification and management of the processes employed within an organization and particularly the interactions between such processes (ISO 9000:2000). See also *process management*.

Process capability The limits within which a tool or process operates, based upon minimum variability as governed by the prevailing circumstances (ASQ, 1998). A statistical measure of the inherent process variability for a given characteristic (QP, 2002).

Process capability index The value of the tolerance specified for the characteristic divided by the process capability (QP, 2002).

Process control The methodology for keeping a process within boundaries; minimizing the variation of a process (QP, 2002).

Process improvement The application of the plan-do-check-act (PDCA) cycle philosophy to a process to produce improvement and better meet the needs and expectations of customers (QP, 2002).

Process management The pertinent techniques and tools applied to a process to implement and improve process effectiveness, hold the gains and ensure process integrity in fulfilling customer requirements (QP, 2002).

Process map A type of flowchart depicting the steps in a process, with identification of responsibility for each step and the key measure (QP, 2002).

Process owner The person who coordinates the various functions and work activities at all levels of a process, has the authority or ability to make changes in the process as required and manages the entire process to ensure performance effectiveness (QP, 2002).

Producer's risk The probability that an acceptable lot of product will be rejected under a specified sampling plan (ASQ, 1998). See also: *consumer's risk*.

Product Result of a process (ISO 9000:2000).
The ISO 9000 standard recognizes four generic product categories: services, software, hardware, and processed materials. Food products are examples of processed materials, while food service is an example of a service.

Project Unique process consisting of a set of coordinated and controlled activities with start and finish dates, undertaken to achieve an objective conforming to specific requirements including the constraints of time, cost and resources (ISO 9000:2000).

Qualification process Process to demonstrate the ability to fulfil specified requirements (ISO 9000:2000).

Quality Degree to which a set of inherent characteristics fulfils requirements (ISO 9000:2000). The totality of characteristics of a product or service that bear on its ability to satisfy stated and implied needs (ASQ, 1998).

Quality assurance (QA) Part of quality management focused on providing confidence that quality requirements will be fulfilled (ISO 9000:2000). All those planned or systematic actions necessary to provide adequate confidence that a product or service will satisfy given needs (ASQ, 1998). All the planned and systematic activities implemented within a quality system that can be demonstrated to provide confidence a product or service will fulfill requirements for quality (QP, 2002). See also: *quality control.*

Quality audit See: *audit.*

Quality characteristic Inherent characteristic of a product, process or system related to a requirement (ISO 9000:2000). See also: *characteristic.*

Quality control (QC) Part of quality management focused on fulfilling quality requirements (ISO 9000:2000). The operational techniques and activities that sustain a quality of product or service that will satisfy given needs; also the use of such techniques and activities. (ASQ, 1998) The operational techniques and activities used to fulfill requirements for quality (QP, 2002). See also: *quality assurance.*

Quality improvement Part of quality management focused on increasing the ability to fulfill quality requirements (ISO 9000:2000). See also: *continual improvement.*

Quality management Coordinated activities to direct and control an organization with regard to quality (ISO 9000:2000).The totality of functions involved in the determination and achievement of quality (ASQ, 1998). See also: *total quality management.*

Quality management principles The ISO 9000:2000 standard identifies eight quality management principles that can be used by an organization's top management in order to lead the organization toward improved performance. These principles are: customer focus, leadership, involvement of people, process approach, system approach to management, continual improvement, factual approach to decision making, mutually beneficial supplier relationships.

Quality management system Management system to direct and control an organization with regard to quality (ISO 9000:2000). A formalized system that documents the structure, responsibilities and procedures required to achieve effective quality management (QP, 2002).

Quality manual Document specifying the quality management system of an organization (ISO 9000:2000).

Quality objective Something sought, or aimed for, related to quality (ISO 9000:2000).

Quality plan Document specifying which procedures and associated resources shall be applied by whom and when to a specific project, product, process, or contract (ISO 9000:2000) A document or set of documents that describe the standards, quality practices, resources and processes pertinent to a specific product, service or project (QP, 2002).

Quality planning Part of quality management focused on setting quality objectives and specifying necessary operational processes and related resources to fulfill the quality objectives (ISO 9000:2000).

Quality policy Overall intentions and direction of an organization related to quality as formally expressed by top management (ISO 9000:2000). An organization's general statement of its beliefs about quality, how quality will come about, and what is expected to result (QP, 2002).

Random cause A cause of variation due to chance and not assignable to any factor (QP, 2002).

Record Document stating results achieved or providing evidence of activities performed (ISO 9000:2000).

Regrade Alteration of the grade of a nonconforming product in order to make it conform to requirements differing from the initial ones (ISO 9000:2000).

Release Permission to proceed to the next stage of a process (ISO 9000:2000).

Reliability The ability of an item to perform a required function under stated conditions for a standard period of time (ASQ, 1998). The probability of a product performing its intended function under stated conditions without failure for a given period of time (QP, 2002).

Repair Action taken on a nonconforming product to make it acceptable for the intended use (ISO 9000:2000). See also: *rework*.

Requirement Need or expectation that is stated, generally implied or obligatory (ISO 9000:2000). Examples include product requirements,

Food Quality Assurance: Principles and Practices

customer requirements, and quality management system requirements.

Review Activity undertaken to determine the suitability, adequacy, and effectiveness of the subject matter to achieve established objectives
(ISO 9000:2000).

Rework Action on a nonconforming product to make it conform to the
requirements (ISO 9000:2000).

Scatter diagram A graph relating two process variables. Useful for investi-
gating the relationship (if any) between these variables. It is one of
the seven tools of quality (ASQ, 1998). A graphical technique to
analyze the relationship between two variables (QP, 2002).

Scrap Action taken on a nonconforming product to preclude its originally
intended use (ISO 9000:2000).

Severity The seriousness of a hazard's effect(s) (NACMCF, 1997).

Seven tools of quality Tools that help organizations understand their processes in order to improve them. The tools are: cause-and-effect
diagram, check sheet, control chart, flowchart, histogram, Pareto
chart, and scatter diagram (QP, 2002).

Shewhart cycle See: *plan-do-check-act cycle.*

Six Sigma A methodology that provides businesses with the tools to improve the capability of their business processes. This increase in
performance and decrease in process variation leads to defect reduction and improvement in profits, employee morale, and product quality (QP, 2002).

Six Sigma quality A term generally used to indicate that a process is well
controlled (i.e., six sigma from the centerline in a control chart) (QP,
2002).

Special causes Causes of variation that arise because of special circumstances. They are not an inherent part of a process. Special causes are
also referred to as assignable causes (QP, 2002). See also: *common
causes.*

Specification Document stating requirements (ISO 9000:2000) A document
that states the requirements to which a given product or service
must conform (QP, 2002).

Statistical process control (SPC) The application of statistical techniques to
control a process (QP, 2002). See also: *statistical quality control.*

Statistical quality control (SQC) The application of statistical techniques to
control quality, includes statistical process control (QP, 2002).

Step A point, procedure, operation, or stage in the food chain including raw materials, from primary production to final consumption (Codex Alimentarius, 1997). A point, procedure, operation, or stage in the food system from primary production to final consumption (NACMCF, 1997).

Supplier Organization or person that provides a product (ISO 9000:2000). An organization that provides a product or service to the customer (ASQ, 1998). A source of materials, service, or information input provided to a process (QP, 2002).

Supply chain The series of suppliers relating to a given process (QP, 2002).

Surveillance Monitoring or observation to verify whether an item or activity conforms to specified requirements (ASQ, 1998). The continual monitoring of a process; a type of periodic assessment or audit conducted to determine whether a process continues to perform to a predetermined standard (QP, 2002).

System Set of interrelated or interacting elements (ISO 9000:2000). A group of interdependent processes and people that together perform a common mission (QP, 2002).

System approach to management This is one of the eight quality management principles recognized by ISO 9000:2000. This principle states: "Identifying, understanding and managing interrelated processes as a system contributes to the organization's effectiveness and efficiency in achieving its objectives."

Technical expert (audit) Person who provides specific knowledge or expertise on the subject to be audited (ISO 9000:2000).

Test Determination of one or more characteristics according to a procedure (ISO 9000:2000).

Testing A means of determining the capability of an item to meet specified requirements by subjecting the item to a set of physical, chemical, environmental or operating actions and conditions (ASQ, 1998).

Top management Person or group of people who directs and controls an organization at the highest level (ISO 9000:2000).

Total quality control (TQC) A system that integrates quality development, maintenance and improvement of the parts of an organization. It helps a company to economically manufacture its product and deliver its services (QP, 2002).

Total quality management (TQM) A term initially coined by the U.S. Naval Air Systems Command to describe its Japanese-style management approach to quality improvement. TQM is a management approach to long-term success through customer satisfaction. It is based on the participation of all members of an organization in improving processes, products, services, and the culture in which they work. The methods for implementing this approach are found in the teachings of such quality leaders as Philip B. Crosby, W. Edwards Deming, Armand V. Feigenbaum, Kaoru Ishikawa, and Joseph M. Juran (QP, 2002).

Traceability Ability to trace the history, application or location of that which is under consideration (ISO 9000:2000). The ability to trace the history, application, or location of an item and like items or activities by means of recorded identification (ASQ, 1998).

Upper control limit (UCL) Control limit for points above the central line in a control chart (QP, 2002). See also: *lower control limit.*

Validation Confirmation through the provision of objective evidence that the requirements for a specific intended use or application have been fulfilled (ISO 9000:2000). Obtaining evidence that the elements of the HACCP plan are effective (Codex Alimentarius, 1997). That element of verification focused on collecting and evaluating scientific and technical information to determine whether the HACCP plan, when properly implemented, will effectively control the hazards (NACMCF, 1997). The act of confirming that a product or service meets the requirements for which it is intended (QP, 2002). See also: *verification.*

Value added The parts of the process that add worth from the perspective of the external customer (QP, 2000).

Variables (method of) Measurement of quality consisting of measuring and recording the numerical magnitude of a quality characteristic for each of the units in the group under consideration (ASQ, 1998).

Verification Confirmation, through the provision of objective evidence, that specified requirements have been fulfilled. (ISO 9000:2000). The application of methods, procedures, tests, and other evaluations, in addition to monitoring to determine compliance with the HACCP plan (Codex Alimentarius, 1997). Those activities, other than monitoring, that determine the validity of the HACCP plan and that the system is operating according to the plan (NACMCF, 1997). The act of determining whether products and services conform to specific requirements (QP, 2002).

World-class quality A term used to indicate a standard of excellence: best of the best (QP, 2002).

Work environment Set of conditions under which work is performed (ISO 9000:2000).

In food processing, control of the work environment is required as good manufacturing practice and as part of the prerequisite programs of HACCP.

Zero defects A performance standard developed by Philip Crosby (QP, 2002).

1.3 Recognized experts in the quality field

The evolution of quality management principles and practices has resulted from the pioneering work of several recognized experts during the last century and particularly since the 1950s. Among the experts whose contributions are particularly well recognized are:

Philip Crosby introduced the zero defects concept. He has authored several books in the field of quality management (QP, 2002).

W. Edwards Deming developed the 14 points referred to as the 14 quality management practices to help companies increase their quality and productivity. He has authored several books in the field of quality management and is also recognized for his expertise in statistical quality control (QP, 2002).

Harold F. Dodge is recognized for his work with acceptance sampling and standardized inspection procedures. He has contributed to the development of several acceptance sampling concepts, including consumer's risk, producer's risk and average outgoing quality level. Together with Harry G. Romig, he developed the Dodge-Romig sampling tables (QP, 2002).

Armand, V. Feigenbaum, the author of *Total Quality Control*, is recognized for being the first to propose the concept (QP, 2002).

Eugene L. Grant introduced statistical quality control concepts to improve manufacturing production. He has authored and co-authored several books (QP, 2002).

Kaoru Ishikawa is recognized for the development of the cause-and-effect diagram which is also known as the Ishikawa diagram. He is regarded as a pioneer in quality control activities in Japan and has authored several books in the field of quality control and quality management (QP, 2002).

Joseph M. Juran is recognized for his contributions to the quality management; he has authored or co-authored several books on various topics in the field of quality management (QP, 2002).

Harry G. Romig was the first to develop the sampling plans using variable data and the concept of average outgoing quality limit. Together with Harold F. Dodge, he developed the Dodge-Romig sampling tables (QP, 2002).

Walter A. Shewhart is recognized for his pioneering work in bringing together the disciplines of statistics, engineering, and economics. This work is the subject of his *book Economic Control of Quality of Manufactured Product*. He has authored books on statistical quality control and is best known for creating the control chart (QP, 2002).

Genichi Taguchi developed methodology to improve quality and reduce costs, this is referred to as the Taguchi methods. He also developed the concept of quality loss function (QP, 2002).

*References**

ASQ, Food Processing Industry Quality System Guidelines, ASQ Food, Drug, and Cosmetic Division, ASQ Quality Press, Milwaukee, 1998.

Codex Alimentarius, *1997 Basic Texts on Food Hygiene,* Codex Alimentarius Commission, Joint FAO/WHO Food Standards Program, Rome.

ISO 9000:2000 International Standard, 2nd ed., 2002–12–15, *Quality management systems — fundamentals and vocabulary,* ISO, Geneva.

NACMCF, 1997. Hazard Analysis and Critical Control Point Principles and Application Guidelines, adopted in August 1997 by the National Advisory Committee on Microbiological Criteria for Foods, *J. Food Prot.*, 61(9), 1998, 1246–1259.

QP, Quality glossary, *Qual. Prog.* 35(7), American Society for Quality, Milwaukee, July 2002, pp. 43–61.

---

* Definitions were reprinted with permission from sources cited.

chapter two

Overview of food quality and food safety

2.1 Introduction

This chapter provides an overview of the principles and practices identified with safety and quality in the food industry. These principles and practices are based on laws and government regulations, as well as the requirements and expectations of customers and consumers. In addition to the basic need for food quality and food safety activities, operations in the food industry have been influenced by numerous factors since the 1970s, including:

- Consumer expectations relating to various aspects of food (e.g., nutrition, convenience, additives)
- Incidents relating to food safety
- Environmental concerns
- Changes in government regulatory processes
- Traceability in food production and processing
- Technological changes
- Foods derived from biotechnology
- Irradiated foods
- Organic foods
- Economic factors
- Issues relating to international trade
- Food security concerns related to bioterrorism

As a result, there have been substantial changes in food industry operations, including significant developments in food quality and food safety activities.

In its continual efforts to address the food quality and food safety requirements of government, customers, and consumers, and in the face of the challenges mentioned earlier, the food industry has embraced generic quality systems and quality programs similar to those used by other industries. At the same time, it has adopted safety systems and programs that have been developed specifically for use within the food industry. Consequently, food quality and food safety requirements are addressed through the use of systems and programs that include quality management, quality assurance, quality control, the hazard analysis critical control point (HACCP) system, and good manufacturing practices (GMPs). Within a particular food company, the food quality and food safety activities are likely to be covered by some combination of these programs or systems.

2.2   Scope of food quality and food safety

In addressing food quality and food safety, it is important to keep in mind that the term "food" covers any unprocessed, semi-processed, or processed item that is intended to be used as food or drink. This includes any ingredient incorporated into a food or drink, and any substance that comes into direct contact with a food during processing, preparation, or treatment. Therefore, food quality and food safety principles and practices are applied to foods from farm produce and livestock production; manufactured and processed food products for consumers; and all raw materials, ingredients, processing aids, food-contact packaging materials, and food-contact surfaces that are used in the preparation of food and beverage products.

The scope of food quality and food safety covers foods already in the marketplace and new or modified foods. When new or modified foods are developed for the marketplace, quality and safety must be considered at the conception, design, and development stages.

2.3   Responsibility for food quality and food safety

The overall responsibility for food quality and food safety is shared by all segments of the food system, including the various food industry sectors, government regulatory agencies, and consumers in general. The food industry has both the legal and moral responsibility for providing customers and consumers with foods that meet all established quality and safety requirements. Within a food company, overall responsibility for the implementation and effective use of these programs and systems rests with senior management.

Governments worldwide have enacted food laws and regulations designed to ensure that foods are fit for human consumption. Such laws protect consumers from harm resulting from unsafe foods and from deception resulting from misrepresentation or fraud relating to certain established food quality characteristics. Governments have also established various agencies that enforce these food laws and regulations; this legal framework is intended to provide consumers with confidence in the safety and quality of foods.

Within the food supply chain, customers who purchase raw materials, ingredients and food contact packaging materials for manufacture of consumer foods, must ensure that these materials are safe and fit for use. When making purchases, consumers need to be vigilant in their assessment of foods for safety and quality. In particular, customers and consumers must pay attention to the instructions for handling, storage, preparation, and use of foods.

2.4   The distinction between food quality and food safety

While the terms *food quality* and *food safety* are often used interchangeably, it is important for the food industry professional to distinguish between them. Food quality is the extent to which the all the established requirements relating to the characteristics of a food are met. Food safety is the extent to which those requirements relating specifically to characteristics or properties that have the potential to be harmful to health or to cause illness or injury are met. Some food quality characteristics (e.g., counts of total bacteria, coliform bacteria) can be used as indicators of food safety, although they are not considered specifically as food safety characteristics. This distinction between food quality and food safety needs to be made, primarily because of the much greater importance that must be attached to protecting consumers from food-borne illnesses or injuries. A food that does not conform to the food safety requirements automatically does not conform to the food quality requirements. On the other hand, a food can conform to the food safety requirements, but not conform to the other quality requirements.

2.5   Food safety as part of food quality

In the food industry, food safety principles and practices have always been integrated into activities identified within quality assurance or quality control programs, or within quality management systems; therefore, these programs and systems can address both food quality and food safety simultaneously. The more recent use of HACCP systems in some food companies has resulted in a well-defined set of activities that are specifically devoted to food safety. The principles and practices of the HACCP system are similar to those of quality systems and, therefore, the specific activities required by the HACCP system can be integrated within quality systems. A food company that operates with a quality management system can be expected to have an HACCP system that is devoted specifically to food safety as an integral part of its quality management system. A food company that does not operate with the HACCP system must continue to incorporate food safety activities and GMPs within its existing quality program or quality system.

Government agencies that use HACCP-based programs to monitor and enforce food laws and regulations are essentially addressing food safety and fitness for use as human food. The HACCP-based programs do not address some of the quality aspects of food laws and regulations. Nevertheless, it is common for the same government agency to monitor and enforce both the food safety and food quality aspects of those laws and regulations. Examples of HACCP-based programs that are used by government regulatory agencies are the U.S. Food and Drug Administration Seafood HACCP Regulation, and Juice HACCP Regulation; the U.S. Department of Agriculture Pathogen Reduction: HACCP System Regulations; and the HACCP-based Food Safety Enhancement Program of the Canadian Food Inspection Agency.

2.6   Food quality

Food quality, as distinct from food safety (Section 2.8), is the extent to which all the established requirements relating to the characteristics of a food are met. Common examples of quality characteristics of food, excluding the food safety characteristics, are:

- Identity of a food in relation to a standard (e.g., standardized food)
- Declared gross or net quantity (e.g., weight or volume) of a unit of the food or net fill of a food container
- Declared or claimed amount of one or more stated components of a food
- Appearance (e.g., size, shape, color)
- Flavor
- Aroma
- Texture
- Viscosity
- Shelf-life stability
- Fitness for use as human food
- Wholesomeness
- Adulteration
- Packaging
- Labeling

Some of these quality characteristics are covered in food laws and regulations. For instance, failure of a food to meet regulatory requirements relating to a standard of identity, the declared quantity, declared ingredients, or label claims, can be considered as misrepresentation, misbranding, or fraud. The spoilage, deterioration, or decomposition of foods with the absence of any resulting harmful substance that can lead to illness or injury, can be considered as failure to meet food quality requirements based on fitness for human use or wholesomeness criteria. Unacceptable levels of foreign matter or extraneous materials that are not necessarily harmful to health or do not cause injury can also be considered as failure to meet food quality requirements; in the U.S., defect action levels have been established for naturally occurring, unavoidable, extraneous materials in many foods. The Codex Alimentarius defines the term *food suitability* (distinct from food safety) as the assurance that food is acceptable for human consumption according to its intended use; food suitability criteria include fitness for human use, wholesomeness, and extraneous matter.

In addition to the quality requirements established by government regulations, numerous requirements for food quality characteristics are also established by customers and consumers. Purchases of food from a manufacturer or supplier by customers and consumers depend on whether the food meets the quality requirements established by the customer or the expectation of the consumer.

2.7 Systems and programs for food quality

The food industry, like many other industries, has used basic quality control programs, and more complex quality assurance programs and quality management systems, in its efforts to achieve food quality; some food companies use the ISO 9000 Quality Management System Standard. These programs and systems can include components that are devoted specifically to food safety. For instance, GMPs and the HACCP system can be integrated into a food industry, quality management system, or inspection and monitoring of materials, products, and processes for food safety hazards can be part of a quality control program. Chapter 3 is devoted to quality programs and quality systems that are used to achieve food quality and food safety.

2.8   Food safety

Food safety is the assurance that food will not cause harm to the consumer when it is prepared and eaten according to its intended use. All requirements relating to the safety characteristics of a food must be met; there must be no unacceptable health risk associated with a food. The assurance that a food will not cause harm, injury, or illness is determined by: (1) whether all harmful substances present in the food have been eliminated, reduced to an established acceptable level, or prevented from exceeding the acceptable level; and (2) the food has been prepared, handled, and stored under controlled and sanitary conditions in conformance with practices prescribed by government regulations. The harmful substances in foods are food safety hazards (Section 2.14). The prescribed conditions and practices for preparing, handling, and storing food are considered GMPs (Section 2.13).

2.9   Systems and programs for food safety

For decades, the food industry has depended on the use of quality programs based on inspection and testing of food products for hazards, and on GMPs for addressing food safety. Since the late 1980s, there has been widespread use of the HACCP system specifically to achieve food safety; the system addresses food safety primarily on the basis of prevention or elimination of unacceptable hazard levels. The GMPs, which were used to address food safety requirements prior to the use of the HACCP system, have been incorporated into prerequisite programs for the HACCP system. A food company that does not operate with the HACCP system must continue to use the GMPs. Chapter 4 is devoted to GMPs and prerequisite programs for the HACCP system; Chapter 5 is devoted to the HACCP system.

2.10 *Food laws and regulations*

The legal requirements for food safety and food quality have been established by many national governments, with the objective of protecting consumers and ensuring that foods are fit for human consumption. These requirements are contained in food laws and regulations, the scope of which varies from one country to another. In the U.S. and Canada, food laws and regulations govern all aspects of food safety and some aspects of food quality. The food laws and regulations of the U.S. are likely the most extensive of any country. It is essential that food industry professionals be familiar with the laws and regulations that govern their specific industry sectors in their countries.

The legal framework of food laws and regulations of a particular country depends on the overall government regulatory system of that country. In the U.S. and Canada, the federal or national food laws are statements of government policies that cover both the general and specific aspects of adulteration and misbranding of foods, while the food regulations deal with the enforcement of government policies that are embodied in the food laws. These food laws and regulations are intended to ensure that foods do not cause harm, illness, or injury; are not adulterated or misbranded; and are wholesome and fit for human consumption. Food laws and regulations apply to all foods produced domestically, as well as all foods imported into a country; foods cannot be imported if they do not conform to the food laws and regulations of the importing country. Examples of food laws are the U.S. Federal Food, Drug, and Cosmetic Act (FDCA), which is the primary law governing the safety and quality of most foods in the U.S., and Canada's Food and Drugs Act, which is the primary food law in Canada. The U.S. Code of Federal Regulations (CFR) Title 21 and Canada's Food and Drug Regulations are examples of food regulations that address food safety and food quality.

Food laws protect consumers from illnesses and injury by prohibiting the presence of any poisonous or harmful substance in foods that are intended for human consumption. For example, in the U.S., adulterated food is regulated primarily under the FDCA, which covers all aspects of food safety and certain aspects of food quality. In addition, food laws protect consumers from fraud and deception by prohibiting false or misleading information relating to foods For example, in the U.S., misbranded food is prohibited under the FDCA.

2.11   Enforcement of food laws and regulations

The responsibility for enforcing food laws and regulations is assigned to government regulatory agencies. These enforcement activities fall into two categories. First, they include inspection and audit of establishments that process, handle, and store food to ensure that the required sanitary and controlled conditions are followed; audits are used by some regulatory agencies that enforce HACCP-based regulations. Second, they include inspection and analysis of foods for harmful substances to ensure that there is conformance to established limits and tolerances.

Despite efforts of government agencies to enforce food laws and regulations, misbranded foods or foods that cause harm or have the potential to cause harm sometimes enter the food distribution chain or the consumer market. Whenever a misbranded food is detected, a harmful substance or agent is detected in a food, it is determined that there is a likelihood for a harmful substance or agent to be present in a food, or an actual food-borne illness or injury occurs, food companies and government regulatory agencies take the necessary action to protect consumers against these violations. These situations often result in the food being recalled from the marketplace. In addition, if it is determined that adulterated or misbranded food has been produced as a result of negligence on the part of a food company, legal action can be taken against the company.

2.12   Food standards

In addition to food laws and regulations, food standards also establish requirements for the safety and quality of foods; however, unless a food standard is part of food regulations (e.g., standard of identity in the U.S. CFR Title 21), it is not a legal requirement. The Codex Standards are the best examples of food standards. The Codex Alimentarius Commission has the mandate to implement the joint Food and Agricultural Organization (FAO)/ World Health Organization (WHO) Foods Standards Program. This has resulted in the Codex Alimentarius, a collection of standards for food quality, food suitability, and food safety. These food standards have been adopted by countries worldwide and are intended primarily to protect consumers and to facilitate international food trade. They include codes of practice such as The Codex General Principles Of Food Hygiene, standards for maximum residual levels (MRL) for pesticides and for veterinary drugs in foods, and standards for specifications for food additives.

2.13   Food quality, food safety, and good manufacturing practices (GMPs)

Government regulatory agencies have established minimum requirements relating to the sanitary practices and controlled conditions for processing, handling, and storage of foods (e.g., current good manufacturing regulations in U.S. CFR Title 21, Part 110). These requirements are commonly referred to as GMPs, and are some of the basic food quality and food safety activities in food companies. If a food is prepared, handled, or stored under conditions that are unsanitary, or if certain required practices or operations are not followed, the food can be considered to be potentially unsafe, unfit, or unsuitable for consumption. Food companies that operate with the HACCP system, incorporate the GMPs within the HACCP prerequisite programs. The GMPs required for food safety are covered in Chapter 4.

2.14   Food safety and hazards in foods

The safety of a food can be related directly to certain harmful substances that are present in the food; these substances are food safety hazards. Any substance that is reasonably likely to cause harm, injury or illness, when present above an established acceptable level, is a food safety hazard. An unacceptable level of a food safety hazard in a food presents a health risk to the consumer. Food-borne illnesses from food safety hazards occur frequently; each year a relatively large number of deaths attributed to these hazards occur among North American consumers.

There are three recognized categories of food safety hazards: biological hazards, chemical hazards, and physical hazards (Sections 2.16 to 2.18). The origin of these hazards in foods can be from naturally occurring substances or agents in foods, from deterioration or decomposition of foods, or from contamination of the foods with the hazard at various stages of their production, harvesting, storing, processing, distribution, preparation, and utilization. For many hazards, government regulatory agencies have established an acceptable level of the hazard in a food; the Codex Alimentarius has also established acceptable levels of certain hazards as part of its Food Standards Programme. For some hazards, such as pathogenic bacteria (e.g., *Salmonella* spp.), there is zero tolerance; this means that the presence or the detection of the hazard in the food is unacceptable. The strategies used to address hazards in foods include the prevention or elimination of hazards, or the reduction of hazards to acceptable levels. These strategies are employed in the HACCP system.

2.15   Food safety hazards and health risk

For a known food safety hazard, the extent of the harmful effects of the hazard on the health of the consumer is established by risk analysis and by hazard analysis. Risk analysis is usually conducted by a national food or health regulatory agency and addresses a public health concern regarding a particular food safety hazard associated with a sector of the food industry. A risk analysis is comprised of risk assessment, risk management, and risk communication. A primary objective of risk analysis is to establish a national food safety objective for a hazard in a food. The food safety objective for a hazard is the maximum frequency and concentration of a hazard in a food at the time of consumption that provides the appropriate level of protection from the hazard. The food safety objective can be considered as the maximum acceptable level for the hazard in a food.

At the level of production, processing, handling, or storage, a food company performs hazard analysis as part of the development of an HACCP plan for the food. Hazard analysis is the first of the seven HACCP principles, and is performed to determine the health risk associated with a hazard present in a food when it is produced, processed, handled, or stored, according to an established sequence of steps at a particular location. Once a food safety objective for a hazard has been established by risk analysis, it must be considered during the hazard analysis step of HACCP plan development.

2.16 Biological hazards in foods

2.16.1 Pathogenic bacteria

Food-borne pathogenic bacteria are responsible for a large proportion of food poisoning incidents in North America. Therefore, the importance of this group of hazards must be emphasized. More than forty different pathogenic bacteria are known; however, a large proportion of the reported cases of food poisoning can be attributed to the following pathogenic bacteria: *Salmonella* spp., *Eschericha coli 0157:H7, Lysteria monocytogenes, Clostridium perfringens, Clostridium botulinum, Staphylococcus aureus,* and *Campylobacter jejeuni.* Food poisoning from these organisms occur frequently, with symptoms that include headache, muscle pain, nausea, fatigue, chills or fever, stomach or abdominal pain, vomiting, and diarrhea. Numerous severe and fatal illnesses occur as a result of food poisoning from pathogenic bacteria; infants and the elderly are particularly vulnerable. The foods that are commonly involved in these food poisoning incidents include meat and poultry and their products, seafood and seafood products, egg and egg products, milk and dairy products, fruits and vegetables and their products, low-acid canned foods, and water.

2.16.2 Viruses

Foods can be the medium for transmission of certain viruses. Examples of viruses that are known to be food safety hazards are the hepatitis A and E viruses, the Norwalk group of viruses, and rotavirus.

2.16.3 Parasites

Several human parasites can be transmitted by foods. The most common human parasites include parasitic protozoan species (e.g., *Entamoeba histolytica, Giardia lambia, Cryptosporidium parvum*), and parasitic worms (*Ascaris lumbricoides, Taenia solium, Trichinella spiralis*).

2.17 Chemical hazards in foods

2.17.1 Permitted food additives

Government regulations permit numerous chemical and biochemical substances to be added to foods at specified maximum levels. These substances are intended to impart some improved nutritional effect (e.g., vitamin fortification) or some specific technical function (e.g., preservative action, sensory attribute, stabilizing effect, etc.). Permissible food additives with their established levels for use can be found listed in government food regulations (e.g., U.S. CFR Title 21, Canada's Food and Drug Regulations). In addition, the Codex Alimentarius contains specifications of permitted food additives. Although food additives are permitted by government regulations, many can be harmful if they are present in the food at levels above the maximum established, and are therefore, potential chemical hazards. In some instances, a permitted food additive present below the maximum allowable level in a food can be a health hazard for specific segments of the population. For example, sodium bisulfite is a permitted food additive in some foods; however, individuals who are asthmatic could be at risk from foods containing sodium bisulfite. The labels on the containers containing the foods must clearly indicate the presence of the additives for the benefit of individuals who may be at risk from these additives.

2.17.2 Naturally occurring harmful compounds

It is well known that many foods contain as their normal or inherent components naturally occurring substances that can be harmful if they are present in excess of certain levels; examples are oxalate in rhubarb, alkaloids in potatoes, toxins in mushrooms and in shellfish. In the U.S., the FDCA considers foods containing these naturally occurring substances to be adulterated only if the harmful substance is present in sufficient quantity that is likely to cause illness.

2.17.3 Unavoidable contaminants

Some foods can contain naturally occurring harmful substances that are not normal or inherent components of the foods. These substances are considered unavoidable contaminants in the food and cannot be removed through processing or manufacturing practices; examples are aflatoxins from molds in peanuts and in some cereals. If the normal level of a naturally occurring harmful substance in a food is increased to an unsafe level as a result of mishandling of the food or by any other action, then the harmful substance can be considered as an added harmful substance.

2.17.4   Agricultural residues

Agricultural residues are a group of residual chemical or biochemical substances found in foods and are directly attributable to certain substances that have been approved for use in the production of crops and livestock for food. They include residues of permitted pesticides, herbicides, fungicides, drugs, hormones, and antibiotics. Some of these residues are considered as added harmful substances attributable to human actions and are regulated by governments. In the U.S., these residues are regulated under several laws including the FDCA. The Codex Alimentarius establishes maximum residual levels (MRL) for various harmful pesticides and veterinary drugs.

2.17.5   Industrial contaminants

Several harmful chemicals that enter the environment as a result of industrial activity have been shown to be present in foods. These substances include heavy metals (lead, mercury, arsenic), organo-chlorinated compounds such as polychlorinated biphenyls (PCBs), and are considered as industrial or environmental contaminants. In the U.S., the CFR Title 21 considers PCBs as unavoidable environmental contaminants because of their widespread occurrence in the environment, and provides tolerances for PCB residues in several foods (e.g., milk, dairy products, poultry, eggs, etc.).

2.17.6   Chemical residues

In food processing operations, some chemical compounds that are not permitted substances in food are used during certain operations and care must be taken to prevent unintentional contamination. These substances include chemical compounds used for cleaning and sanitizing food contact surfaces of processing, handling, and storage equipment, and for lubricating certain parts of food processing equipment.

2.17.7   Prohibited chemicals

No chemical substance is permitted for use in a food unless it meets all of the requirements that are covered in the applicable food laws and regulations. In addition, in the U.S. CFR Title 21, some chemical substances are specifically prohibited from direct addition to food or from indirect addition to food through food contact surfaces.

2.17.8   Food allergens

Certain foods are known to contain inherent components that cause serious immunological, allergic responses in a relatively small proportion of food consumers. These foods are entirely safe for most consumers who are not sensitive to the allergens. The following foods and some of their products are generally considered to be the most common food allergens: peanuts, soybeans, milk, eggs, fish, crustacea, tree nuts, and wheat. Some other foods (e.g., sesame seeds) are also known to cause allergenicity occasionally.

In addition, sulfites (including bisulfites and metabisulfites) used as ingredients in certain foods can produce nonimmunological allergic reactions in certain sensitive individuals.

2.18   Physical hazards in foods

Physical hazards include organic or inorganic substances, commonly referred to as foreign objects, foreign matter, or extraneous materials. Hard and sharp physical hazards are of particular concern. Depending on their size and dimensions, hard and sharp physical hazards can cause injury to the mouth or teeth, or can cause serious injuries if swallowed. In addition, some physical hazards, depending on their size, shape, and texture, have the potential to cause choking if swallowed. Physical hazards in foods can be particularly harmful to infants.

Certain hard and sharp foreign objects that are natural components of food (e.g., prune, date or olive pits; fish bones nutshells) are not considered physical hazards since it is expected that the consumer will be aware that these objects are natural components of the foods. However, if the food carries a label stating that the hard and sharp object has been removed(e.g., pitted prunes), the presence of the hard and sharp object in the food represents a hazard, since it is not expected by the consumer.

The common hazards considered as avoidable physical hazards in foods include broken glass, pieces of hard or soft plastic materials, stones, pieces of metal, pieces of wood, and personal articles.

2.18.1   Broken glass

In a food plant, the common potential sources of broken glass include light bulbs, glass containers, and gauges with glass covers. Every effort must be taken to protect or eliminate these sources of broken glass, and to protect food from contamination with this hazard. In addition, many foods are packaged, distributed and sold in glass containers. For these foods, the glass packaging itself can be a source of broken glass.

2.18.2   Plastic

Both hard and soft plastic foreign objects are sometimes found in foods. In some food plants, some utensils and tools used for cleaning of equipment are made from hard plastic material; this type of plastic can become brittle from use over an extended period of time, and pieces can adulterate foods. The common sources of soft plastic foreign objects in food are plastic material used for packaging food and gloves used by employees who handle food.

2.18.3 Metal pieces

The most common sources of metal pieces in a food plant are food processing equipment, metallic cleaning tools, and equipment maintenance activities. In many food plants, magnets are used to eliminate some metals from foods, and metal detectors are used to detect the presence of metals in foods.

2.18.4 Wood pieces

The most common sources of wood pieces in a food plant are wood structures and wood pallets. The presence of these sources should be avoided whenever possible in food processing and production.

2.18.5 Stones

Many plant foods and particularly field crops such as peas and beans can contain small stones that become incorporated with the foods during harvesting. In addition, in food processing plants, a common source of stones is concrete structures, particularly concrete floors.

2.18.6 Personal articles

A variety of personal articles can become foreign objects in foods, resulting from unintentional adulteration by employees during preparation, handling, processing, and packaging. Personal articles that have been found in foods include jewelry, pens or pencils or their parts, Band-Aids, and ear plugs.

2.19 Other food safety concerns

In addition to the known food safety hazards that can be classified as biological, chemical and physical hazards, there are several other specific food safety concerns. These include concerns relating to the safety of foods from biotechnology and particularly from genetically modified organisms, the safety of irradiated foods, and the safety of some herbal supplements and botanical products. The safety of these foods, like all other foods, is covered by food laws and regulations (Section 2.10). In the food industry, the established food safety principles and practices must also be applied to these food safety concerns.

References

Codex Alimentarius, FAO/WHO Food Standards, www.codexalimentarius.net/.
Code of Federal Regulations, Title 21, Food and Drugs, National Archives and Records Administration, www.access.gpo.gov/cgi-bin/cfrassemble.cgi?title=200221.
Federal Food, Drug, And Cosmetic Act, U.S. Food and Drug Administration, www.fda.gov/opacom/laws/fdcact/fdctoc.htm.

Food and Drug Act, Food and Drug Regulations, Health Canada, hc-gc.ca/food-aliment/friia-raaii/food_drugs-aliments_drogues/act-loi/e.index.html.

FSEP Implementation Manual, Canadian Food Inspection Agency Food Safety Enhancement Program, www.inspection.gc.ca/english/ppc/psps/haccp/manu/manue.shtml.

chapter three

Quality programs and quality systems for the food industry

3.1 Introduction

This chapter is devoted to the principles and practices associated with quality programs and quality systems that can be applied to operations in the food industry. In general, these quality programs and systems are quality control, quality assurance, and quality management. They are generic in nature and are widely used by business organizations not only in the food industry, but in all industry sectors, as well as in some public sector organizations. These programs and systems differ in their scope of activities and the complexity of their structure or framework. Quality control programs are basic quality programs, and quality management systems are more complex types of quality systems. In the food industry, the objective of these programs and systems is to achieve the food quality and food safety requirements.

An explanation of the structure and contents of the quality management systems of the *ISO 9001:2000* and *ISO 9004:2000* standards comprise a substantial part of this chapter. The *ISO 9001* standard is the most widely used quality standard in the business world. The 2000 edition of the *ISO 9000* family of standards incorporates quality management principles and fundamentals for quality management systems. In this chapter, the *ISO 9000* family of standards is used as a basis for describing quality principles and practices for the food industry. The reader will find it useful to refer to the *ISO 9001:2000* and *ISO 9004:2000* standards.

3.2 The distinction between quality programs and quality systems

In the food industry, a quality program is an activity or set of activities performed to ensure that the food quality and food safety requirements of a food are fulfilled. Food quality requirements are established by laws and regulations and by customers and consumers. A food industry quality system is an integrated set of documented food quality and food safety activities, with clearly established inter-relationships among the various activities. The objective of a quality system is to provide a food company with the capability to produce a food that fulfills all quality and safety requirements. Quality control programs are common examples of quality programs; quality assurance systems and quality management systems are examples of quality systems. Both quality programs and quality systems are used extensively in the food industry.

3.3   Quality control programs

Quality control program activities consist of inspecting, testing, and monitoring associated with raw materials control, process control, and finished products control. The main objective of food industry quality control programs is to determine whether the quality and safety requirements are fulfilled by detecting whether unacceptable levels of hazards or defects exist in foods. If an unacceptable level of a hazard or defect is detected, the food might be repaired or reworked to remove the hazard or defect so that it fulfils the requirements, or it might be rejected entirely and scrapped.

The goal of a food company's quality control program is to ensure that all requirements are fulfilled so that only safe foods of acceptable quality are sent to its customers or to consumers. In companies that operate with quality systems, the quality control activities are integrated into the quality systems. For example, in the *ISO 9001:2000* quality management system, quality control activities are included in clause *7 product realization* (Section 3.14.7) and clause *8 measurement, analysis, and improvement* (Section 3.14.8).

3.4   Quality assurance systems

Quality assurance systems in the food industry are much more extensive in scope than quality control programs. They include the inspection, testing, and monitoring activities of quality control programs, along with additional activities that are devoted to prevention of food safety hazards and quality defects. The activities are integrated and interrelated to form a system. Quality assurance systems are intended to provide confidence to a food company's management, its customers and to government regulatory agencies that the company is capable of meeting the food quality and food safety requirements. These quality systems include documents that describe operations and activities that directly relate to food quality and safety. An example of a quality assurance system is the *ISO 9001:1994* quality assurance system standard, which was replaced by the *ISO 9001:2000* quality management system standard. In companies that operate with quality management systems, the quality assurance activities are integrated into the quality management systems.

3.5   Quality management systems

Quality management systems are elaborate management systems that can be used by any organization to develop and achieve its quality objectives. Quality management systems include quality planning and improvement activities, in addition to quality control and assurance activities. These systems are intended to provide a company with the capability to meet all quality requirements. The best example of a quality management system is the *ISO 9001:2000 Quality management system — requirements* standard (Sections 3.12 and 3.14). In the past, the terms *total quality control* and *companywide quality control* were occasionally used in the same context as quality management systems.

3.6   Total quality management

During the mid-1980s, the term *total quality management* (TQM) was introduced in North America. The term was associated with the management approach to quality improvement used in Japan for achieving long-term success. The TQM approach embodies both management principles and quality concepts, including customer focus, empowerment of people, leadership, strategic planning, improvement, and process management. These principles and concepts evolved during the second half of the twentieth century with substantial contributions from several recognized experts in the field of quality management. Of these contributions, the most widely recognized are the 14 points for quality management proposed by W. Edwards Deming. During the 1980s and 1990s many North American businesses adopted the TQM approach and developed the framework for its use in their quality management systems, with the objective of achieving competitive advantage in the global marketplace.

3.7   Recognition programs and excellence models

Governments in some developed countries have instituted programs aimed at recognizing organizations that use the TQM approach in their quality management systems. The objective of these recognition programs is to promote and foster the use of quality management principles, concepts, and practices within organizations, and particularly for achieving competitive advantage in the global marketplace. These recognition programs are based on an established framework for TQM and are essentially excellence models for organizations that use TQM to achieve world-class quality. In North America, the U.S. Malcolm Baldrige National Quality Award (MBNQA) administered by the National Institute for Standards and Technology (NIST),

Table 3.1 Seven Categories of Quality Management Activities of the U.S. Malcolm Baldrige National Quality Award (MBQNA) and Eight ISO Quality Management Principles

MBNQA Categories	ISO Quality management principles
Leadership	Customer focus
Strategic planning	Leadership
Customer and market focus	Involvement of people
Information and analysis	Process approach
Human resource focus	System approach to management
Process management	Continual improvement
Results	Factual approach to decision making
	Mutually beneficial supplier relationships

is the best known example of a government recognition program for organizations that use TQM. In Japan, the recognition program is the Deming Prize. In the MBNQA program, organizations that compete for the award are evaluated on the basis of their performance in seven categories of activities. These categories are listed in Table 3.1; for comparison, the eight quality management principles recognized by ISO in *ISO 9004:2000* (Section 3.11.1) are also given.

3.8 Quality system standards

A quality system standard is a document that describes the requirements of a quality system. The *ISO 9001:2000* quality management system standard is the recognized international quality system standard. Many countries have formally adopted this international standard as their national quality system standard. Prior to the adoption of the international quality system standard, some countries had developed their own national quality system standards. In addition, some industry sectors have developed sector-specific quality system standards. In some instances, these sector-specific quality system standards are based on the *ISO 9000* quality standard; an example is the QS 9000 standard of the North American automotive industry.

3.9 The ISO 9000 quality system standards

The *ISO 9000* quality system standards were developed by the International Organization for Standardization (ISO) for use by any organization that needs to develop, implement and operate with a quality management system. The *ISO 9000* quality system standards have had considerable impact on the evolution of quality activities on a global scale since the first set of standards were issued in 1987. Estimates indicate that the *ISO 9000* quality system standards are used by more than half a million organizations worldwide.

3.10 The evolution of the ISO 9000 quality system standards

The *ISO 9000* quality system standards, which were introduced in 1987, were revised in 1994 and again in 2000. The objective of these periodic revisions is to satisfy the needs of the standards' users. In the 2000 revision, quality management principles and fundamentals of quality management systems were formally recognized and incorporated into the *ISO 9000* family of standards. The 2000 revision, which was the first major revision of the standards, resulted in the following three standards:

- *ISO 9000:2000* Quality management systems—fundamentals and vocabulary
- *ISO 9001:2000* Quality management systems—requirements
- *ISO 9004:2000* Quality management systems—guidelines for performance improvements

The two standards *ISO 9001:2000* and *ISO 9004:2000* have been referred to as a consistent pair, indicating that they are compatible with each other. *ISO 9001:2000* defines minimum requirements for an organization that seeks to have its quality management system recognized by a formal registration process. *ISO 9004:2000* is not used for registration but provides guidance for performance improvement of a quality management system. The scope of this guidance extends much beyond the minimum requirements of *ISO 9001:2000*.

The 2000 edition of the *ISO 9000* family of standards cancelled and replaced the 1994 edition of several of the standards, including:

- ISO 8402:1994 Quality management and quality assurance—vocabulary
- ISO 9000—1:1994 Quality management and quality assurance standards–part 1: guidelines for selection and use
- ISO 9001:1994 Quality systems—model for quality assurance in design, development, production, installation and servicing
- ISO 9002:1994 Quality systems—model for quality assurance in production, installation and servicing
- ISO 9003: 1994 Quality systems—model for quality assurance in final inspection and test
- *ISO 9004–1: 1994* Quality management and quality system elements—part 1: guidelines

In addition to these changes in the 2000 revisions, several documents in the *ISO 9000* family of documents were retired, while some others were revised.

The 1994 revisions were considered minor revisions with little change in the structure of the initial 1987 version of the three standards *ISO 9001*, *ISO 9002*, and *ISO 9003*, which were models for quality assurance. However, in the 2000 revisions the structure of the standards was considerably modified. This led to cancellation of the *ISO 9001*, *ISO 9002*, and *ISO 9003* quality assurance system standards and their replacement by the single *9001:2000* quality management system standard. The *ISO 9001:2000* standard, the requirements of which are described in five clauses (Table 3.2), was substantially restructured when compared with the *ISO 9001:1994* standard in which the requirements were described in 20 elements.

3.11 ISO 9000:2000 quality management systems — fundamentals and vocabulary

The information in the *ISO 9000:2000* standard is extremely important for an understanding of the basics of quality management systems. The standard introduces a set of eight quality management principles and describes a set of 12 fundamentals, which serve as the basis for the ISO quality management systems. The quality management principles are embodied in the quality management system fundamentals. The standard also provides definitions of technical terms (Chapter 1).

3.11.1 ISO 9000:2000 quality management principles

The eight quality management principles recognized in the *ISO 9000:2000* standard are:

- **Customer focus:** The success of an organization is dependent on the extent to which the current and future needs and requirements of customers are known, understood, and met. Therefore, the organization must devote substantial effort toward meeting the requirements of its customers and should strive to exceed the expectations of its customers.

- **Leadership:** Senior managers must establish the direction and the objectives of the organization, and must ensure that the conditions exist for achieving these objectives. This leadership is required to ensure there is a common purpose for everyone within the organization.

- **Involvement of people:** The involvement of everyone within the organization is essential to achieving the objectives. Personnel must have the responsibility, authority, abilities, and skills, and the tools required for them to contribute fully to the organization.

- **Process approach:** An organization's activities are performed more effectively and efficiently when managed as a process. Therefore, an organization should use the process approach to manage its activities.

- **System approach to management:** The processes carried out by an organization should be identified, understood, and managed as an interrelated set of processes that form a complete system. The effectiveness and efficiency of the entire organization can be enhanced by adopting this system approach.

- **Continual improvement:** One of the ongoing objectives of an organization should be the improvement of its performance on a continual basis.

- **Factual approach to decision making:** An organization should compile and analyze information for use in decision making.

- **Mutually beneficial supplier relationships:** An organization can benefit from developing relationships with its suppliers; these relationships serve to enhance the performance of both the organization and its suppliers.

3.11.2   ISO 9000:2000 fundamentals of quality management systems:

The *ISO 9000* standard recognizes the following 12 fundamentals, which are the basis for the contents of the *ISO 9001:2000* and *ISO 9004:2000* quality management system standards. These fundamentals, which incorporate the eight quality management principles (Section 3.11.1), are:

- **Rationale for quality management systems:** A quality management system can provide benefits to an organization. In general, these benefits include:
  - Assist in enhancing the satisfaction of the organization's customers
  - Provide a framework for continual improvement in the organization
  - Provide confidence to the organization and its customers that the organization has the capability to provide products that meet the requirements of customers, regulatory agencies and the organization.

- **Requirements for quality management systems and requirements for products:** The *ISO 9000* standards distinguish between the requirements of quality management systems and requirements of products. The *ISO 9001:2000* standard provides generic quality management system requirements that are applicable to any organization but does not provide requirements for an organization's products. An organization, its customers, and government regulatory agencies establish requirements for products; these requirements are also part of the quality management system.

- **Quality management systems approach:** In the development, implementation, maintenance, and improvement of its quality management system, an organization needs to adopt an approach in which certain specified activities should be undertaken; the standard identifies these activities.

- **The process approach:** The process approach is described as systematic identification and management of an organization's processes and the interactions between these processes. This approach should be used to manage an organization.

- **Quality policy and quality objectives:** An organization's quality policy and quality objectives can provide a focus for the direction of the organization. The quality policy should provide a framework for establishing the quality objectives, which should be consistent with the quality policy.

- **Role of top management within the quality management system:** An organization's top management, through the use of quality management principles (Section 3.11.1), and its leadership and actions can create an environment for the involvement of its people and for effective operation of the organization's quality management system.

- **Documentation:** Documentation is an essential feature of an organization's quality management system. Various types of documents are needed in a quality management system, each should serve a particular function.

- **Evaluating quality management systems:** An organization's quality management system should be assessed by evaluating the various processes within the system, by auditing the system, and by top management's review of the system. An organization should also carry out self-assessment of its activities and performance.

- **Continual improvement:** An organization's quality management system should include activities that are devoted to continually improving the system with the objective of enhancing the satisfaction of its customers and other interested parties.

- **Role of statistical techniques:** An organization should use statistical techniques to understand and solve problems such as variability, for continual improvement of its effectiveness and efficiency, and in making decisions.

- **Quality management systems and other management system focuses:** An organization's quality management system can be integrated with other management systems (e.g., financial management system, environmental management system, employee health and safety management system). The quality objectives of the quality management system can complement the objectives of the other management systems.

- **Relationship between quality management systems and excellence models:** The approach of *ISO 9000:2000* family of standards has many similarities to those of excellence models (Section 3.7). However, the *ISO 9000* standards provide quality management system requirements (*ISO 9001*) and guidance for performance improvement (*ISO 9004*), while the excellence models provide assessment criteria for comparing an organization's performance against the performance of other organizations.

3.12 ISO 9001:2000 quality management systems–requirements

The *ISO 9001* standard is the most widely used standard among the *ISO 9000* family of standards. It is used by companies seeking to have their quality management systems recognized through an independent registration process. This standard establishes requirements for an organization's quality management system. It is designed for use by an organization that aims to enhance the satisfaction of its customers and needs to demonstrate that it has the ability to provide products that meet requirements of customers, regulatory agencies, and the organization. The contents of the *ISO 9001* standard are covered in Section 3.14.

3.13 ISO 9004:2000 quality management systems–guidelines for performance improvements

The *ISO 9004* standard provides guidelines, in contrast to the *ISO 9001* standard, which specifies requirements. The *ISO 9004* standard is designed for use by an organization that seeks to move beyond the requirements of *ISO 9001*, to improve the performance of the organization, and to satisfy its customers and its other interested parties. The *ISO 9004* standard covers both the effectiveness and efficiency of an organization's quality management system. By contrast, the *ISO 9000* standard covers the effectiveness of an organization's quality management system. The contents of the *ISO 9004* standard are covered in Section 3.14.

3.14 ISO 9001:2000 and ISO 9004:2000 standards

The remainder of this chapter covers the quality management system requirements in the *ISO 9001:2000* standard and the guidelines for performance improvements in the *ISO 9004:2000* standard. These requirements

Table 3.2 Clauses of the ISO 9001:2000 Quality Management
System—Requirements and ISO 9004 Quality Management System—Guidelines
for Performance Improvements.

ISO 9001:2000	ISO 9004:2000
1. Scope	**1. Scope**
1.1 General	
1.2 Application	
2. Normative reference	**2. Normative reference**
3. Terms and definitions	**3. Terms and definitions**
4. Quality management system	**4. Quality management system**
4.1 General requirements	4.1 Managing systems and processes
4.2 Documentation requirements	4.2 Documentation
4.2.1 General	
4.2.2 Quality manual	
4.2.3 Control of documents	
4.2.4 Control of quality records	
	4.3 Use of quality management principles
5. Management responsibility	**5. Management responsibility**
5.1 Management commitment	5.1 General guidance
	5.1.1 Introduction
	5.1.2 Issues to be considered
5.2 Customer focus	5.2 Needs and expectations of interested parties
	5.2.1 General
	5.2.2 Needs and expectations
	5.2.3 Statutory and regulatory requirements
5.3 Quality policy	5.3 Quality policy
5.4 Planning	5.4 Planning
5.4.1 Quality objectives	5.4.1 Quality objectives
5.4.2 Quality management system planning	5.4.2 Quality planning
5.5 Responsibility	5.5 Responsibility, authority, and communication
5.5.1 Responsibility and authority	5.5.1 Responsibility and authority
5.5.2 Management representative	5.5.2 Management representative
5.5.3 Internal communication	5.5.3 Internal communication
5.6 Management review	5.6 Management review
5.6.1 General	5.6.1 General
5.6.2 Review input	5.6.2 Review input
5.6.3 Review output	5.6.3 Review output
6. Resource management	**6. Resource management**
6.1 Provision of resources	6.1 General guidance
	6.1.1 Introduction
	6.1.2 Issues to be considered

-- continued

Table 3.2 (continued) Clauses of the ISO 9001:2000 Quality Management System—Requirements and ISO 9004 Quality Management System—Guidelines for Performance Improvements.

ISO 9001:2000	ISO 9004:2000
6.2 Human resources	6.2 People
6.2.1 General	6.2.1 Involvement of people
6.2.2 Competence, awareness and training	6.2.2 Competence awareness and training
	6.2.2.1 Competence
	6.2.2.2 Awareness and training
6.3 Infrastructure	6.3 Infrastructure
6.4 Work environment	6.4 Work environment
	6.5 Information
	6.6 Suppliers and partnerships
	6.7 Natural resources
	6.8 Financial resources
7. Product realization	**7. Product realization**
7.1 Planning pf product realization	7.1 General guidance
	7.1.1 Introduction
	7.1.2 Issues to be considered
	7.1.2.1 General
	7.1.2.2 Process inputs, outputs, and review
	7.1.2.3 Product and process validation and changes
7.2 Customer related process	7.2 Processes related to interested parties
7.2.1 Determination of requirements related to the product	
7.2.2 Review of requirements related to the product	
7.2.3 Customer communication	
7.3 Design and development	7.3 Design and development
7.3.1 Design and development planning	7.3.1 General guidance
7.3.2 Design and development inputs	7.3.2 Design and development input and output
7.3.3 Design and development outputs	7.3.3 Design and development review
7.3.4 Design and development review	
7.3.5 Design and development verification	
7.3.6 Design and development and validation	
7.3.7 Control of design and development changes	

-- *continued*

Table 3.2 (continued) Clauses of the ISO 9001:2000 Quality Management System—Requirements and ISO 9004 Quality Management System—Guidelines for Performance Improvements.

ISO 9001:2000	ISO 9004:2000
7.4 Purchasing	7.4 Purchasing
7.4.1 Purchasing process	7.4.1 Purchasing processes
7.4.2 Purchasing information	7.4.2 Supplier control process
7.4.3 Verification of purchased product	
7.5 Production and service provision	7.5 Production and service operations
7.5.1 Control of production and service provision	7.5.1 Operation and realization
7.5.2 Validation of processes for production and service provision	7.5.2 Identification and traceability
7.5.3 Identification and traceability	7.5.3 Customer property
7.5.4 Customer property	
7.5.5 Preservation of product	7.5.4 Preservation of product
7.6 Control of monitoring and measuring devices	7.6 Control of measuring and monitoring devices
8. Measurement, analysis, and improvement	**8. Measurement, analysis, and improvement**
8.1 General	8.1 General guidance
	8.1.1 Introduction
	8.1.2 Issues to be considered
8.2 Monitoring and measurement	8.2 Measurement and monitoring
8.2.1 Customer satisfaction	8.2.1 Measurement and monitoring of system performance
	8.2.1.1 General
	8.2.1.2 Measurement and monitoring of customer satisfaction
	8.2.1.3 Internal audit
	8.2.1.4 Financial measures
	8.2.1.5 Self-assessment
8.2.2 Internal audits	8.2.2 Measurement and monitoring of processes
8.2.3 Monitoring and measurement of processes	8.2.3 Measurement and monitoring of product
8.2.4 Monitoring and measurement of product	8.2.4 Measurement and monitoring the satisfaction of interested parties

-- *continued*

Table 3.2 (continued) Clauses of the ISO 9001:2000 Quality Management System—Requirements and ISO 9004 Quality Management System—Guidelines for Performance Improvements.

ISO 9001:2000	ISO 9004:2000
8.3 Control of nonconforming product	8.3 Control of nonconformity
	8.3.1 General
	8.3.2 Nonconformity review and disposition
8.4 Analysis of data	8.4 Analysis of data
8.5 Improvement	8.5 Improvement
8.5.1 Continual improvement	8.5.1 General
8.5.2 Corrective action	8.5.2 Corrective action
8.5.3 Preventive action	8.5.3 Loss prevention
	8.5.4 Continual improvement of the organization

Source: *ISO 9001:2000, ISO 9004:2000*. With permission.

and guidelines are referred to by the identical numbers and titles used in the standard and are italicized for easy identification. The quality management system requirements and guidelines are applicable to the operations of food companies.

The standard defines the minimum requirements that must be implemented in order for the quality management system to be recognized. An organization that would like to move beyond the requirements of *ISO 9001:2000* can use the guidelines given in *ISO 9004:2000*. Both standards contain the following eight clauses:

1. *Scope*
2. *Normative reference*
3. *Terms and definitions*
4. *Quality management system*
5. *Management responsibility*
6. *Resource management*
7. *Product realization*
8. *Measurement, analysis, and improvement*

Clauses 1 to 3 provide general remarks on the standards. Clauses 4 to 8 describe the requirements (*ISO 9001*) and guidelines (*ISO 9004*). Clause *4 quality management system* defines certain general requirements, which apply to all activities, while clauses *5 management responsibility, 6 resource management, 7 product realization*, and *8 measurement, analysis, and improvement* are considered the four major blocks of activities. Table 3.2 lists and compares the clauses of *ISO 9001:2000* and *ISO 9004:2000*.

3.14.1 ISO 9001:2000 and ISO 9004:2000 clause 1 scope

This clause defines the scope and the objectives, use and application of the standards.

3.14.1.1   ISO 9001 clause 1 scope

Clause *1.1 General* declares that the standard is an international standard that specifies requirements for a quality management system and identifies two particular situations for using the standard:

- When an organization needs to demonstrate that it is has the ability to provide products that meet the requirements of its customers and the regulatory requirements that apply to its products and operations
- When an organization's objective is to enhance the satisfaction of its customers through continual improvement

The scope of the standard is limited to products intended for, or required by, an organization's customers.

Clause *1.2 Application* covers the application of the standard. The requirements of the standard are generic and can be applied regardless of type or size of an organization or type of products. This clause recognizes that because of the generic nature of the requirements of the standard, it is possible that certain requirements may not be applicable to some organizations, depending on the nature of the organization and its type of operations and products. In these situations, an organization can exclude a requirement of the standard that is considered to be not applicable; however, only requirements contained in Clause 7 (*Product Realization*) of the standard can be considered for exclusion.

The standard can be used by any food company to achieve the food quality and food safety requirements of its products.

3.14.1.2   ISO 9004 clause 1 scope

This clause describes the relationship of this standard to the *ISO 9001* standard. The following comparisons are made between the *ISO 9001* and *ISO 9004* standards:

- *ISO 9004* provides guidelines that go beyond the requirements of *ISO 9001*
- The *ISO 9001* objectives relating to customer satisfaction and product quality are extended in *ISO 9004* to include satisfaction of interested parties and the performance of an organization
- Unlike the *ISO 9001* standard, which consists of requirements and can be used for certification or for contractual purposes, the *ISO 9004* standard consists of guidelines and is not intended to be used for certification or for contractual purposes.

3.14.2   *ISO 9001:2000 and ISO 9004:2000 clause 2 normative reference*

This clause is identical for both standards. It identifies the reference document that is applicable to both standards. This reference document is the standard *ISO 9000:2000 quality management systems — fundamentals and vocabulary* (Section 3.11). In the use of the standards, the definitions of terms in the *ISO 9000* standard are the recognized definitions.

3.14.3   *ISO 9001:2000 and ISO 9004:2000 clause 3 terms and definitions*

This clause is identical for both standards. It makes specific reference to *the ISO 9000 standard* as the source of the terms and definitions that are used in *ISO 9001* and *ISO 9004*. In addition, it clarifies the definitions of the terms *product* and *supplier*, and the supply chain relationship involving supplier, organization, and customer. A supplier provides product to an organization, while a customer receives product from an organization.

3.14.4   *ISO 9001:2000 and ISO 9004:2000 clause 4 quality management system*

ISO 9001:2000 Clause 4 provides general requirements and documentation requirements that apply to the entire quality management system, including the other clauses of the standard. *ISO 9004:2000* Clause 4 provides guidelines for managing systems and processes, documentation, and use of quality management principles. These guidelines apply to the entire quality management system, including the other clauses of the standard.

3.14.4.1   *ISO 9001 clause 4.1 general requirements*

A quality management system must be established, documented, implemented, and maintained by the organization, and the effectiveness of the system must be continually improved. This must be done in accordance with the requirements of the standard. In addition, the process approach of the standard is emphasized in this clause, by requiring that the organization address the following points:

- The organization must identify and manage the processes to be included in the quality management system. The standard specifically mentions processes for communication, product realization, customer-related issues, purchasing, production and service provision, monitoring, measurement, analysis, and improvement.
- The sequence and interaction of these processes must be determined.
- The criteria and methods required to ensure the effective operation and control of the processes must be determined.

- The required resources and information must be made available to operate and monitor these processes.
- The processes must be monitored, measured, and analyzed.
- The actions required to obtain expected results, and for continual improvement of the processes, must be implemented.

If an organization outsources any of the processes that affect the quality of its products, the outsourced processes must be included in the organization's quality management system.

3.14.4.2   ISO 9004 clause 4.1 managing systems and processes

This clause recognizes the importance of quality management in the management of an organization. It associates success with the implementation and maintenance of a management system that aims to continually improve both the effectiveness and efficiency of the organization's performance, and the organization's systems and processes in relation to the needs of interested parties. The standard recommends that a customer-oriented organization should be established by the organization's top management and proposes that this should be achieved by:

- Defining the organization's systems and processes in such a manner that they can be managed, and both their effectiveness and efficiency can be improved
- Ensuring that there is effective and efficient operation and control of the organization's processes, and the measures and data that are used to determine its performance is satisfactory.

The clause provides examples of activities that can result in a customer-oriented organization, including the organization's use of processes that lead to performance improvement, the continuous use of data and information from processes, and the use of self-assessments and management reviews to evaluate the improvement of processes.

3.14.4.3   ISO 9001 clause 4.2 documentation requirements

The documentation requirements for an organization's quality management system are given in four clauses.

Clause 4.2.1 General identifies five categories of documents that must be included in an organization's quality management system:

- Documented statements of the organization's quality policy and quality objectives
- The organization's quality manual (see clause 4.2.2)

- Procedures that must be documented by the organization based on the requirements of the standard. Documented procedures are required specifically by the following six clauses of the standard:
  - 4.2.3. *Control of documents*
  - 4.2.4. *Control of records*
  - 8.2.2. *Internal audit*
  - 8.3.   *Control of nonconforming product*
  - 8.5.2. *Corrective action*
  - 8.5.3. *Preventive action*

- Documents that the organization must have in order to plan, operate, and control its processes
- Records that must be maintained:
  - Records from the organization's management reviews (*5.6 Management review, 5.6.1 General*)
  - Records of education, training, skills, and experience for personnel who perform work that affects the quality of the organization's product (*6.2.2 Competence, awareness, and training*)
  - Records to demonstrate that the organization's product realization processes and the products resulting from these processes meet the requirements (*7.1 Planning of product realization*)
  - Records of the results of review of requirements of the organization's products, and of the actions resulting from this review (*7.2.2 Review of requirements related to product*)
  - Records of design and development inputs relating to requirements of the organization's products (*7.3.2 Design and development inputs*)
  - Records of the results of design and development reviews, and any actions resulting from these reviews (*7.3.4 Design and development review*)
  - Records of the results of design and development verification, and any actions from this verification (*7.3.5 Design and development verification*)
  - Records of the results of design and development validation, and any actions from this validation (*7.3.6 Design and development validation*)
  - Records of review of design and development changes, and any actions from this review (*7.3.3 Control of design and development changes*)
  - Records of results of evaluations of the organization's suppliers, and any actions resulting from these evaluations (*7.4.1 Purchasing process*)
  - Records to demonstrate the validation of the organization's processes from which output cannot be verified by subsequent monitoring or measurement (*7.5.2 Validation of processes for production and service provision*)

- Records for unique identification of the organization's product, where traceability is required *(7.5.3 Identification and traceability)*
- For property that is provided to the organization by its customer, the records of customer property that is lost, damaged, or unsuitable for use *(7.5.4 Customer property)*
- Records of the basis used for calibration or verification of measuring equipment where no international or national standards exist *(7.6 Control of monitoring and measuring devices)*
- Records of the validity of previous measuring results when the measuring equipment is found not to conform to requirements *(7.6 Control of monitoring and measuring devices)*
- Records of results of calibration and verification of monitoring and measuring devices used to provide evidence that the organization's products meet requirements *(7.6 Control of monitoring and measuring devices)*
- Records of the organization's internal audits *(8.2.2 Internal audit)*
- Records of conformity of the organization's product based on the acceptance criteria and the personnel responsible for the release of the organization's product *(8.2.4 Monitoring and measurement of product)*
- Records of nonconformities of the organization's products and subsequent actions relating to the nonconforming products *(8.3 Control of nonconforming product)*
- Records of the results of corrective action taken by the organization to eliminate the cause of nonconformities in order to prevent recurrence *(8.5.2 Corrective action)*
- Records of the results of preventive action taken by the organization to eliminate the causes of potential nonconformities in order to prevent their occurrence *(8.5.3 Preventive action)*.

Clause 4.2.2 *Quality manual* defines the requirements for the organization's quality manual. At a minimum, the quality manual must contain:

- The scope of the organization's quality management system, along with information on and justification of which clauses (if any) of the standard are not included; exclusions are permitted only from clause 7
- The organization's documented procedures which are included in the quality management system, or some reference to these documented procedures
- The interaction between the various processes of the organization's quality management system

Clause 4.2.3 *Control of documents* requires that a documented procedure must be established to control the categories of documents identified in clause 4.2.1, except quality records (clause 4.2.4 addresses control of quality records specifically). The documented procedure for control of documents in an organization's quality management system must specify the controls required for the following:

- Approving the documents to ensure that they are adequate, before they are issued
- Reviewing and updating of existing documents when this becomes necessary and the re-approval of updated documents
- Ensuring that the changes are identified when documents are updated, and the status of the revised documents are also identified
- Ensuring that correct versions of documents are available at locations where use of the documents is required
- Ensuring that the documents are always legible and can be readily identified
- Ensuring that any documents that originate from outside the organization are identified and controlled with regard to distribution within the organization
- Preventing the use of documents that have become obsolete, and appropriately identifying these documents if they are retained by the organization

The requirements in this clause can be applied to the documents in a food company's HACCP system, including SOPs and HACCP prerequisite programs (Chapter 4), and documents required by *HACCP Principle 7 Establish record–keeping procedures* (Chapter 5).

Clause 4.2.4 *Control of quality records* requires that a documented procedure must be established specifically for the control of the quality records that an organization is required to keep. These records serve as evidence that there is conformity to requirements of the organization's quality management system, as well as evidence of the system's effective operation. The documented procedure for control of these quality records must specify the controls that are required so that the records are:

- Identified in relation to the purpose that they serve
- Available and stored at specified locations, so they can be retrieved readily
- Protected from loss or damage and maintained in a legible manner
- Retained for a defined period of time
- Disposed of after they are no longer required

The requirements in this clause can be applied to the records in a food company's program for GMPs or HACCP system, including records required by HACCP prerequisite programs (Chapter 4) and by *HACCP Principle 7 Establish record–keeping procedures* (Chapter 5).

3.14.4.4   *ISO 9004 clause 4.2 documentation*

This clause provides guidance on the documentation that should be established, maintained and controlled by an organization, beyond that required by *ISO 9001:2000*. The following are examples of points that need to be considered:

- Contractual requirements (e.g., from customers and other interested parties)
- Statutory and regulatory requirements
- Applicable standards (e.g., international, national, regional, industry sector)
- Other internal and external sources of information

This clause also provides examples of criteria that can be used for evaluation of the effectiveness and efficiency of the organization documentation activities.

3.14.4.5   *ISO 9004 clause 4.3 use of quality management principles*

Eight quality management principles (Section 3.11.1) have been integrated into the contents of the *ISO 9004* standard; the *ISO 9001* standard does not have a clause corresponding to this clause of *ISO 9004*. The quality management principles are intended for use by an organization's top management in order to improve its performance. The standard recognizes that benefits will result from the successful use of these eight quality management principles; including improved monetary return, the creation of value, and increased organizational stability.

3.14.5   *ISO 9001:2000 and ISO 9004:2000 clause 5 management responsibility*

The six *management responsibility* clauses of *ISO 9001* define the responsibilities of top management in establishing, implementing, maintaining, and continually improving the organization's quality management system.

The six *management responsibility* clauses of *ISO 9004* are more extensive than those of *ISO 9001*, They provides guidance for an organization's top management, for its management, and for the organization.

3.14.5.1   *ISO 9001 clause 5.1 management commitment*

The top management of an organization must demonstrate its commitment to developing, implementing, and continually improving the organization's quality management system through the following actions:

- Communicate to the entire organization the importance of meeting customer requirements, and statutory and regulatory requirements

- Establish the organization's quality policy (see clause *5.3 Quality policy*)
- Ensure that the organization's quality objectives are established (see clause *5.4.1 Quality objectives*)
- Conduct management reviews (see clause *5.6 Management review*)
- Ensure that resources are available in the organization for the activities required by the quality management system (see clause *6.1 Provision of resources*)

This clause is particularly applicable to food companies. The top management of a food company is responsible for the safety of the food that is produced at a food establishment. In addition, top management's commitment is essential for the success of a food company's HACCP system for food safety (Chapter 5).

3.14.5.2   ISO 9004 clause 5.1 general guidance

Clause *5.1.1 Introduction* provides guidance on the role of management's leadership and its active involvement in the organization's quality management system, which is essential for establishing and maintaining an effective and efficient system, and for increasing customer satisfaction, in order to achieve benefits. Several examples of actions are given. Top management's responsibilities should also include identifying the methods to be used for measuring the organization's performance in order to determine whether objectives are being achieved. Proposed methods of measurement include:

- Using the organization's financial information
- Using external information, including benchmarking and evaluation by an independent third-party
- Measuring the satisfaction of the organization's employees and other interested employees

Clause *5.1.2 Issues to be considered* reinforces the points in clause *4.3 Use of quality management principles*. It recommends that the organization's top management make use of the eight quality management principles in developing, implementing, and managing the quality management system. It proposes that top management demonstrate leadership and commitment in activities such as:

- Understanding the current and future needs, expectations, and requirements of the organization's customers
- Increasing awareness, motivation, and involvement of the organization's employees
- Planning for the future of the organization and for managing change, including innovative or breakthrough changes to processes within the organization

- Identifying support processes that affect the organization's product realization
- Ensuring that the organization's processes operate as an effective and efficient network

3.14.5.3　*ISO 9001 clause 5.2 customer focus*

This clause requires that top management ensure that the organization determines and fulfils the requirements of its customers, with the aim of enhancing customer satisfaction. Specific details relating to customer requirements and customer satisfaction are given in other clauses of the standard (*7.2.1 Determination of requirements related to products, 8.2.1 Customer satisfaction*).

3.14.5.4　*ISO 9004 clause 5.2 needs and expectations of interested parties*

There is no clause in *ISO 9001* that corresponds to clause *5.2* of *ISO 9004*, which addresses the needs and expectations of interested parties of an organization, including its customer. The guidance provided goes much further than the requirements of *ISO 9001*, which address only customer requirements and customer satisfaction.

Clause *5.2.1 General* identifies the interested parties that have needs and expectations of the organization. In addition to the organization's customers and employees, interested parties include: end-users of the organization's products; owners, investors, and shareholders; the organization's suppliers and partners; the community and the general public that are affected by the organization or its products.

Clause *5.2.2 Needs and expectations* provides specific guidance for management's responsibility for (a) understanding and satisfying the needs and expectations of the organization's present and potential customers and end-users, including both their current and future needs and expectations, and (b) understanding and considering the needs of the organization's other interested parties. Guidance is provided to the organization's management for understanding and meeting the needs of interested parties in general, and for understanding and satisfying the needs and expectations of customers and end-users in particular. Interested parties other than customers and end-users include:

- The organization's employees: Management should identify the needs and expectations of the organization's employees for purposes of recognition, work satisfaction, and personal development.
- The organization's owners, investors, and shareholders: To satisfy the needs and expectations of its owners, investors and shareholders, top management should consider the organization's financial results.
- The organization's suppliers and partners: Management should consider developing partnerships with the organization's suppliers.

- The organization's community and society: Management should consider the impact of the organization's activities on the public's health and safety, on the environment, and on the society in general.

Clause 5.3.2 *Statutory and regulatory requirements* provides guidance relating to the organization's knowledge of statutory and regulatory requirements that pertain to its products, processes, and activities.

3.14.5.5 *ISO 9001 clause 5.3 quality policy*

This clause specifies the responsibility of top management to ensure that the organization's documented quality policy satisfies the following requirements:

- The quality policy is appropriate for the purpose of the organization.
- Top management is committed to complying with the requirements of the quality management system, and to the continual improvement of its effectiveness.
- The quality policy provides a framework for establishing and reviewing the organization's quality objectives.
- The quality policy is communicated and understood within the organization.
- The quality policy is reviewed for its continuing suitability.

3.14.5.6 *ISO 9004 clause 5.3 quality policy*

This clause recommends the following as top management's responsibility for an organization's quality policy:

- The quality policy should be used by top management in its efforts to improve the performance of the organization.
- The quality policy should be an important and equal part of the organization's overall policies.
- Top management should consider the needs and expectations of all interested parties, including customers, when establishing the organization's quality policy.

3.14.5.7 *ISO 9001 clause 5.4 planning*

Clause 5.4.1 *Quality objectives* specifies the responsibility of top management to ensure that the organization's quality objectives are established and documented, and that the quality objectives are measurable and are consistent with the organization's quality policy. Quality objectives that are needed to meet requirements of products must also be included (see clause *7.1 Planning for product realization*).

Clause 5.4.2 *Quality management system planning* specifies top management's responsibility for the planning activities that must be established in the organization's quality management system. The following two points must be addressed:

- Planning must be carried out in order that all the requirements of the quality management system, as specified by the standard, will be established and all the organization's quality objectives will be achieved.
- When changes to the organization's quality management system are planned and implemented, all the requirements specified by the standard must be maintained.

3.14.5.8   ISO 9004 clause 5.4 planning

Clause 5.4.1 *Quality objectives* proposes that the strategic planning and quality policy established by an organization's management should be the basis for its quality objectives. These quality objectives should lead to improvement of the organization's performance.

Clause 5.4.2 *Quality planning* proposes that an organization's management should take responsibility for quality planning, which should focus on meeting the organization's quality objectives. Examples of inputs and outputs of quality planning are provided.

3.14.5.9   ISO 9001 clause 5.5 responsibility, authority and communication

Clause 5.5.1 *Responsibility and authority* specifies that top management is responsible for defining the responsibilities and authorities with respect to the various processes within the organization's quality management system, and the interrelation between these responsibilities and authorities. The defined responsibilities and authorities and their interrelation must be communicated within the organization.

Clause 5.5.2 *Management representative* requires that top management appoint a member of the organization's management who must have the responsibility and authority for quality management system activities which include:

- Establishing, implementing, and maintaining the processes that are required for the organization's quality management system
- Reporting on the performance of the quality management system, and the need for any improvement to the system, to the organization's top management.
- Ensuring that awareness of requirements of the organization's customers is promoted throughout the organization.

Clause 5.5.3 *Internal communication* specifies top management's responsibility to ensure that processes for communication are established within the organization, and that the effectiveness of the organization's quality management system is communicated within the organization.

3.14.5.10 ISO 9004 clause 5.5 responsibility, authority and communication

Clause 5.5.1 *Responsibility and authority* proposes that top management should define responsibility and authority within the organization and communicate this throughout the organization. Personnel throughout the organization should be given responsibilities and authority so that they can contribute to the organization's quality objectives.

Clause 5.5.2 *Management representative* proposes that an organization's top management should appoint a management representative who is given responsibility for managing, monitoring, operating, and coordinating the organization's quality management system. The management representative should have the responsibility of communicating with the organization's customers and its interested parties, and for reporting to top management on matters relating the organization's quality management system.

Clause 5.5.3 *Internal communication* proposes that an organization's management should define and implement an effective and efficient process for communicating the organization's quality policy, its requirements, its objectives, and its accomplishments throughout the organization. The clause also recommends that organization management should actively encourage feedback and communications from its personnel, and provides examples of internal communication activities.

3.15.5.11 ISO 9001 clause 5.6 management review

Clause 5.6.1 *General* specifies top management's responsibility for conducting reviews of the organization's quality management system at established intervals, to ensure that the system continues to be suitable, adequate and effective. Records of these reviews must be kept as evidence that they have been done as required. Reviews must evaluate improvement of the organization's quality management system, and the need to make changes to the organization's existing quality policy, quality objectives, and to the system.

Clause 5.6.2 *Review input* specifies the information that must be used when management reviews of the organization's quality management system are conducted. The information includes the following:

- Results of various types of audits of the organization's quality management system (e.g., internal audits, customer audits, third-party audits)
- Information obtained from the organization's customers
- Performance of the organization's processes and conformity of the organization's products to requirements

- Information on the organization's preventive actions and corrective actions
- Any actions that were identified for follow-up at previous management reviews
- The effect of any changes that are planned to take place in the organization's existing quality management system
- Recommendations for the improvement of the effectiveness and efficiency of the organization's quality management system

Clause 5.6.3 *Review output* specifies decisions and actions that result from management review of the organization's quality management system; these must include the following:

- Improvement actions toward the effectiveness of the organization's quality management system and the processes in the system
- Improvement actions toward the organization's products as they relate to requirements of the organization's customers
- Decisions regarding the need for resources within the organization

3.14.5.12   *ISO 9004 clause 5.6 management review*

Clause 5.6.1 General proposes that an organization's top management should develop management review activities that go beyond verifying the effectiveness and efficiency of the organization's quality management system. It recommends a management review process that extends to the entire organization and provides opportunities for exchange of new ideas, open discussion, and leadership input.

Clause 5.6.2 *Review input* provides examples of review inputs that an organization should consider when evaluating the efficiency and effectiveness of its quality management system as part of its management review activities.

Clause 5.6.3 *Review output* proposes that by extending management review activities beyond verification of its quality management system, the output of the reviews can by used by an organization's top management as inputs for improvement processes. Examples of review outputs that can enhance efficiency are provided.

3.14.6 *ISO 9001:2000 and ISO 9004:2000 clause 6 resource management*

Clause 6 provides information related to identifying, providing, and managing various types of resources in the organization. *ISO 9001* clause 6 addresses requirements and *ISO 9004* clause 6 provides guidelines.

3.14.6.1 ISO 9001 clause 6.1 provision of resources

This clause specifies general requirements relating to resources. The organization must determine the resources required and provide these resources for two activities:

- Implementing, maintaining, and continually improving the quality management system
- Enhancing customer satisfaction by meeting customer requirements

3.14.6.2 ISO 9004 clause 6.1 general guidance

Clause 6.1.1 Introduction proposes that the organization's top management ensure that the resources required for implementing strategy and achieving objectives are identified and available. This includes the resources required for operating and improving the organization's quality management system and the satisfaction of its customers and other interested parties. Examples of resources that may be required by the organization are people, infrastructure, work environment, information, suppliers and partners, natural resources, and financial resources.

Clause 6.1.2 Issues to be considered provides guidance on resources that should be considered for improving an organization's performance. These include: tangible resources such as support facilities, intangible resources such as intellectual property, management leadership skills, employee competence, information management and technology, and use of natural resources. Management of these resources includes making them available at the time they are needed and when they can be most effective and efficient, and planning for future resource needs.

3.14.6.3 ISO 9001 clause 6.2 human resources

Clause 6.2.1 General notes that the personnel whose work affects the quality of product must be competent in their field of work. This competence is determined on the basis of the education, training, skills, and experience.

Clause 6.2.2 Competence, awareness, and training specifies that an organization is required to carry out the following activities:

- Determine the competence required for personnel whose work affects the quality of the organization's products
- Provide training to personnel in order that the required competence can be attained
- Evaluate the training provided to determine whether it is effective for acquiring the required competence
- Ensure that personnel are aware of the relevance of their work to product quality, and of their contribution to the achievement of the organization's quality objectives
- Maintain records of the education, training, skills, and experience of personnel whose work affects product quality

This clause is particularly applicable to food companies. Training a food company's employees is essential for achievement of food safety, and is required as part of GMPs, HACCP prerequisite programs, (Chapter 4) and HACCP systems (Chapter 5).

3.14.6.4   ISO 9004 clause 6.2 people

The guidelines in this clause are given in the following 2 clauses: 6.2.1 Involvement of people, 6.2.2 Competence, awareness and training.

Clause 6.2.1 *Involvement of people* proposes that the organization's management should have the involvement and support of its people in efforts to improve the effectiveness and efficiency of the organization and its quality management system. Management should encourage the involvement and development of its people. Numerous examples of how this can be done are given, including providing ongoing training, defining responsibilities and authorities, recognizing and rewarding performance, and ensuring effective teamwork.

Clause 6.2.2 *Competence, awareness, and training* provides guidelines in two separate clauses:

- **Clause 6.2.2.1 *Competence*** proposes that the organization's management should ensure that the necessary employee competence is available for the organization to operate effectively and efficiently. Management should consider the organization's present and expected competence needs.

- **Clause 6.2.2.2 *Awareness and training*** provides guidelines for education and personnel. It recognizes that the objective of training and education is to provide the organization's personnel with knowledge and skills that, together with their experience, improve their competence.

3.14.6.5   ISO 9001:2000 clause 6.3 infrastructure

This clause specifies that an organization must determine, provide and maintain the infrastructure that is required to achieve products that conform to requirements. An organization's infrastructure includes buildings, workspace, utilities, processing equipment, and any services related to infrastructure such as transportation or the communications network.

This clause is particularly applicable to food companies. A food company's premises, facilities and equipment are covered by requirements of GMPs and HACCP prerequisite programs (Chapter 4).

3.14.6.6   ISO 9004:2000 clause 6.3 infrastructure

This clause provides guidelines relating to the infrastructure that is necessary for an organization to realize its products, and to address the needs and expectations of interested parties. The infrastructure includes the organization's premises, workspace, tools, equipment, support services, information and communication technology, and transport services. The organization should have a process for defining the infrastructure needed to achieve effective and efficient product realization.

3.14.6.7   *ISO 9001:2000 clause 6.4 work environment*

An organization must determine the work environment that is needed in order for it to achieve products that conform to requirements, and must manage this work environment.

This clause is particularly applicable to food companies. A food company's work environment is covered by requirements of GMPs and HACCP prerequisite programs (Chapter 4).

3.14.6.8   *ISO 9004:2000 clause 6.4 work environment*

This clause provides guidelines relating to the work environment that is needed to influence the motivation, satisfaction, and performance of an organization's personnel in order to enhance the performance of the organization. It proposes that the organization's management should ensure that this type of work environment exists. Examples of points to be considered in the creation of a suitable work environment are provided.

3.14.6.9   *ISO 9004 clause 6.5 information*

An organization's management should treat data as a fundamental resource for conversion into information, and for the continual development of the organization's knowledge. This knowledge is essential for making factual decisions and can also stimulate innovation. Examples of points to be considered by an organization in the management of its information are provided.

ISO 9001 does not have a clause that corresponds to this clause.

3.14.6.10   *ISO 9004 clause 6.6 suppliers and partnerships*

An organization should establish relationships with its suppliers and partners. These relationships should promote and facilitate communication among an organization and its suppliers and partners with the aim of mutually improving the effectiveness and efficiency of processes that create value. Examples of opportunities for an organization to increase value through working with its suppliers and partners are provided.

ISO 9001 does not have a clause that corresponds to this clause.

3.14.6.11   *ISO 9004 clause 6.7 natural resources*

An organization should consider whether its performance can be influenced by the availability of certain natural resources which are often out of its direct control. The organization should have contingency plans to ensure the availability of these resources or replacement resources in order to prevent or minimize negative effects on its performance.

ISO 9001 does not have a clause that corresponds to this clause.

3.14.6.12   *ISO 9004 clause 6.8 financial resources*

Activities for determining the need for financial resources and the sources of financial resources should be included in resource management activities. An organization's management should (a) plan, make available, and control the financial resources necessary to implement and maintain an effective and efficient quality management system and to achieve the organization's objectives, and (b) consider the development of innovative financial methods to support and encourage improvement of the organization's performance.

This clause recognizes that improving the effectiveness and efficiency of an organization's quality management system can positively influence the financial results of the organization. Examples of these improvements are reducing process and product failures, reducing waste in materials and time, reducing cost of compensation to customers under product guarantees and warrantees, and reducing costs of lost customers and lost markets.

ISO 9001 does not have a clause that corresponds to this clause.

3.14.7   *ISO 9001:2000 and ISO 9004:2000 clause 7 product realization*

The requirements of *ISO 9001:2000* clause 7 *Product realization* are given in six clauses.

The guidelines of *ISO 9004:2000* clause 7 *Product realization* are given in six clauses.

3.14.7.1   *ISO 9001 clause 7.1 planning of product realization*

This clause requires an organization to plan and develop the processes that are needed for product realization. This planning must be consistent with the requirements of the other processes that are also included in the organization's quality management system. As part of planning for realization of the product, the organization must determine the following:

- Its quality objectives and requirements for its products
- The need to establish processes, create documents, and provide resources that are specific to its products
- Activities required by the organization for verification, validation, monitoring, inspection, and test for its products, and the criteria for acceptance of its products
- The records that the organization must maintain as evidence that both its product realization processes and the resulting products conform to requirements.

This clause is particularly applicable to food companies that operate with the HACCP system for addressing the safety of its products (Chapter 5).

3.14.7.2   ISO 9004 clause 7.1 general guidance

Clause 7.1.1 *Introduction* is a statement about processes in relation to product realization. It also proposes that an organization's top management should ensure the effectiveness and efficiency of product realization processes and support processes and their associated networks so that the organization has the capability to satisfy interested parties. Management should define the required outputs of the organization's processes, and should identify the inputs and activities required for the effective and efficient achievement of these outputs.

Clause 7.1.2 *Issues to be considered* identifies specific points relating to the organization's processes for product realization. These points include:

- Defining the inputs of processes and determining the activities, actions, and resources required for processes in order to achieve the desired outputs
- Continual improvement of processes in order to improve the effectiveness and efficiency of the organization's quality management system and its performance
- Documentation of the processes
- The role of the organization's people in its processes
- Benefits from improvement of the efficiency and effectiveness of the organization's processes.

Clause 7.1.3 *Managing processes:* The guidelines of this clause are covered in the following 3 clauses:

- **Clause 7.1.3.1 *General*** proposes that an organization's management should identify processes that are needed to realize products that satisfy the requirements of its customers and its other interested parties. To do this, the organization's management should consider the following points relating to its processes: associated processes and their desired outputs, process steps, activities, flows, control measures, training needs, equipment, methods, information, materials, and other resources. Examples of support process are provided. The clause also proposes that an operating plan should be defined to manage the organization's processes.

- **Clause 7.1.3.2 *Process inputs, outputs and review*** proposes that an organization should identify significant or critical features of products and processes in order to develop an effective and efficient plan for controlling and monitoring the activities within its processes. Examples of input issues are provided. The organization should record process outputs and evaluate the outputs against input requirements and acceptance criteria, taking into account the needs and expectations of its customers and its other interested parties. Management should undertake periodic reviews of the performance of its processes. Examples of points to be considered in this review are provided.

- **Clause 7.1.3.3 *Product and process validation and changes*** proposes that management should carry out product validation activities to demonstrate that its products meet the needs and expectations of its customers and its interested parties. Validation of its processes should also be carried out at appropriate intervals to ensure that the impact of any changes in the processes can be determined and addressed. Examples of activities for validation of products and processes and points that should be addressed are provided. The clause also addresses (a) control of changes to processes and related issues and (b) risk assessment associated with potential failures or faults in its processes.

3.14.7.3   ISO 9001 clause 7.2 customer-related processes

Clause 7.2.1 *Determination of requirements related to the product* explains that an organization must determine the following:

- All product-related requirements that are specified by its customers, including requirements for delivery of its products and post-delivery requirements
- All necessary product requirements for its specified or intended use that are not stated by its customer
- All statutory, regulatory, or other requirements relating to its products
- Any additional requirements determined by the organization

This clause is particularly applicable to food companies that operate with the HACCP system for addressing the safety of their products (Chapter 5).

Clause 7.2.2 *Review of requirements relating to the product* requires that an organization must review all the requirements relating to a product before it makes a commitment to supply products to its customers. This review must ensure the following:

- The requirements relating to products are defined.
- Any differences between the customer's contract or order requirements and requirements that were expressed previously are resolved.
- The organization has the ability to supply to the customer products that meet the defined requirements.
- The organization must keep records of this review and the actions resulting from the review. In cases where the customer does not

provide a documented statement of requirements, the organization must confirm the requirements to the customer before accepting them. In cases where the product requirements of the customer are modified, the organization must ensure that the documents relating to these requirements are changed, and all relevant personnel are made aware of the changes.

Clause 7.2.3 *Customer communication* requires that an organization must determine the effective arrangements for communicating with its customers and must implement these arrangements. This communication must address the following:

- Information on the organization's products
- Inquiries from customers, and handling of customer contracts or orders, including changes to these contracts or orders
- Feedback from customers, including complaints from customers

3.14.7.4   *ISO 9004 clause 7.2 processes related to interested parties*

This clause provides guidelines for an organization's relationships with its customers and with its other interested parties. It proposes that the organization's management should ensure that the organization has defined, implemented, and maintained mutually acceptable processes for communicating effectively and efficiently with its customers and its other interested parties. These processes should serve to ensure adequate understanding of the needs and expectations of the organization's interested parties, and for translation of these needs and expectations into requirements for the organization. These processes should involve the organization's customers and its interested parties and should include the identification and review of relevant information. Examples of information to be reviewed are provided.

3.14.7.5   *ISO 9001 clause 7.3 design and development*

Clause 7.3.1 *Design and development planning* specifies that an organization must plan and control the design and development of its products and, if appropriate, must update the output from this planning as the design and development progresses. During the design and development planning, the organization must determine the following:

- The stages of the design and development of its products
- The review, verification and validation appropriate to each design and development stage
- The responsibilities and authorities for design and development

The organization must manage the interfaces between the different groups in design and development to ensure effective communication and clear assignment of responsibility.

Clause 7.3.2 *Design and development inputs* specifies that an organization must determine the design and development inputs relating to the requirements of products, and must maintain records of these design and development inputs, including the following:

- Functional and performance requirements of the product
- Statutory and regulatory requirements relating to the product
- Information from previous similar designs (if applicable)
- Other requirements that are essential for the design and development of the product

The design and development inputs must be reviewed for adequacy and the requirements of the product must be complete, and must not be ambiguous or conflict with each other.

Clause 7.3.3 *Design and development outputs* specifies that an organization must provide the outputs of design and development of the product in a form that enables verification against the design and development inputs, and must be approved prior to their release. The design and development outputs must:

- Meet the input requirements for design and development of the product
- Provide appropriate information for purchasing, production and for service production
- Contain or refer to the acceptance criteria of the product
- Specify the characteristics of the product that are essential for its safe and proper use.

Clause 7.3.4 *Design and development review* specifies that an organization must perform systematic reviews of the design and development at suitable stages. This review must be done in accordance with arrangements specified in design and development planning and must be done for the following purposes:

- To evaluate the results of the design and development to meet the requirements of the product
- To identify any problems relating to the design and development activities and propose the necessary actions

The participants in these design and development reviews must include representatives from functions concerned with the design and development stages being reviewed. The organization must keep records of the results of these reviews and of any necessary actions resulting from the reviews.

Clause 7.3.5 *Design and development verification* specifies that an organization must carry out design and development verification to ensure that the design and development outputs have met the design and development requirements. This verification must be done in accordance with the arrangements specified in design and development planning. The organization must keep records of the results of this verification and of any actions resulting from the verification.

Clause 7.3.6 *Design and development validation* specifies that an organization must perform design and development validation to ensure that the resulting product is capable of meeting the requirements for the specified application or intended use, when this is known. This validation must be performed in accordance with the arrangements specified in design and development planning, and wherever practical, must be completed prior to the delivery or implementation of the product. The organization must keep records of the results of this validation and of any actions resulting from the validation.

Clause 7.3.7 *Control of design and development changes* states that an organization must identify design and development changes and keep records of these changes. These changes must be reviewed, verified and validated, as appropriate, and approved before they are implemented. The review of design and development changes must include evaluation of the effect of the changes on constituent parts and on product already delivered. The organization must keep records of the results of the review of design and development changes and of any necessary actions resulting from this review.

Note that in the food industry, research and development activities are considered as design and development activities.

3.14.7.6   *ISO 9004 clause 7.3 design and development*

This clause provide guidelines for an organization to consider when it is designing and developing new products and processes in relation to the needs and expectations of customers and other interested parties.

Clause 7.3.1 *General guidance* proposes that the top management of an organization should ensure that the organization has defined, implemented and maintained the necessary design and development processes to respond effectively and efficiently to the needs and expectations of its customers and its other interested parties. Management should ensure that the organization is capable of considering the basic performance and function of products and processes that are being designed and developed, as well as all factors that contribute to meeting the product and process performance expected by its customers and its other interested parties. Examples of the performance characteristics that should be considered are provided. Management should also ensure that steps are taken to identify and mitigate potential risk to the users of the organization's products and processes that are being designed and developed; examples of the tools for risk assessment are provided.

Clause 7.3.2 *Design and development input and output* proposes that the organization should identify process inputs that affect the design and development of its products and processes, and facilitate effective and efficient process performance in order to satisfy the needs and expectations of its customers and its other interested parties. Both the external and internal needs and expectations should translate into input requirements for the design and development processes. Examples of external and internal inputs and other inputs that should be considered are provided. Design and development outputs should include information that will enable verification and validation against the requirements, and should be reviewed against inputs to provide evidence that outputs have effectively and efficiently met the requirements for the process and product being designed and developed. Examples of design and development outputs are provided.

Clause 7.3.3 *Design and development review* proposes that the top management of an organization should ensure that appropriate people are assigned to manage and conduct systematic reviews, at selected points and at completion of design and development, to determine that design and development objectives are achieved. Examples of the topics to be considered at these reviews are provided, and include verification and validation activities.

3.14.7.7 ISO 9001 clause 7.4 purchasing

This clause covers requirements regarding purchases from suppliers.

Clause 7.4.1 *Purchasing process* requires that an organization must ensure that products it purchases conform to specified requirements. The organization must exercise control over both its suppliers of purchased products and the purchased products. The type and extent of this control depends on the effect of the purchased products on subsequent product realization or the organization's final products. The organization must evaluate and select its suppliers based on their ability to supply products in accordance with the organization's requirements, and must keep records of the results of evaluations of its suppliers, and any necessary actions resulting from the evaluations. The organization must establish criteria for selection, evaluation, and re-evaluation of its suppliers.

Clause 7.4.2 *Purchasing information* specifies that the purchasing information an organization provides to its suppliers must describe the products to be purchased, including the following, where appropriate:

- Requirements for the approval of the products, procedures, processes and equipment
- Requirement for qualification of personnel
- Quality management system requirements

The organization must ensure that the specified purchase requirements are adequate before they are communicated to its suppliers.

Clause 7.4.3 *Verification of purchased product* specifies that an organization must establish and implement inspection or other activities necessary for ensuring that the products it purchases meet the specified purchase requirements. In cases where the organization or its customers intend to perform verification at the premises of a supplier, the organization must state in the purchasing information provided to the supplier the intended verification arrangements and the method of release of the purchased product. This clause is particularly applicable to the food industry. Inspection of raw materials, ingredients and packaging materials for defects and food safety hazards are essential practices in the food industry.

3.14.7.8   ISO 9004 clause 7.4 purchasing
This clause provides guidelines for an organization to consider regarding the products it purchases to satisfy its needs and requirements, including those of its interested parties, and the control it should exercise over its suppliers.

Clause 7.4.1 *Purchasing process* proposes that an organization's top management should ensure that effective and efficient purchasing processes are defined and implemented for the evaluation and control of the organization's purchased products. This should be done to ensure that purchased products satisfy the needs and expectations of the organization and of its interested parties. The organization and its suppliers together should develop requirements for the suppliers' processes and specifications for purchased products, and should define the need for records relating to verification of purchased products and nonconforming purchased products. Examples of activities are provided.

Clause 7.4.2 *Supplier control process* proposes that an organization should establish effective and efficient processes to identify potential sources for its purchased materials, to develop its existing suppliers and partners, and to evaluate their ability to supply the products the organization requires. This should be done to ensure the effectiveness and efficiency of the organization's purchasing processes. The organization's management should consider actions necessary to maintain the organization's performance and to satisfy its interested parties in the event of failure by its suppliers. Examples of inputs to the organization's supplier control process are provided.

3.14.7.9   ISO 9001 clause 7.5 product and service provision
Clause 7.5.1 *Control of production and service provision* specifies that an organization must plan its production and service provisions, and carry them out under controlled conditions that include the following, where applicable:

- Availability of information that describes the characteristics of the organization's products

- Availability of work instructions for production and service provision, where necessary
- Use of suitable equipment
- Availability and use of monitoring and measuring devices
- Implementation of monitoring and measurement
- Implementation of activities for release and delivery of products from production and service provision, and for post-delivery activities relating to its products.

This clause is particularly applicable to food companies that carry out process control activities, and those which operate with GMPs and the HACCP system.

Clause 7.5.2 *Validation of processes for production and service provision* specifies that an organization must validate any production and service provision process from which the resulting product cannot be verified by the organization's monitoring or measurement activities. This includes any process for which deficiencies become apparent only after product is in use or the service has been delivered. The validation must demonstrate that these processes have the ability to achieve the planned results. The organization must establish arrangements for these processes; these arrangements must include the following, where applicable:

- Defined criteria for review and approval of these processes
- Approval of the equipment and qualification of the personnel required for these processes
- Use of specific methods and procedures in these processes
- Requirements of records from the activities of these processes
- Revalidation of these processes

Clause 7.5.3 *Identification and traceability* specifies that, where appropriate, an organization must identify its products throughout the product realization process, and must identify the status of its products with respect to monitoring and measurement requirements of the products. If traceability is required, the organization must control and record the unique identification of its products. This clause is particularly applicable to food companies for their product recall procedure which is a HACCP prerequisite program (Chapter 4).

Clause 7.5.4 *Customer property* specifies that an organization must exercise control over the property of its customers, while this property is under the organization's control or is being used by the organization. The organization must identify, verify, protect, and safeguard the customer property that is provided for use or incorporation into the organization's product intended for that customer. The organization must report to its customer, any customer property that is lost, damaged, or found to be unsuitable for use while it is under the organization's control, and must keep records of this.

Clause 7.5.5 *Preservation of product* specifies that an organization must preserve the conformity of its products during internal processing and during delivery to its intended destination. This preservation applies to the organization's products as well as components of its products, and must include identification, handling, packaging, storage, and protection. This clause is particularly applicable to food companies that operate with GMPs and HACCP prerequisite programs (Chapter 4) as part of their HACCP system.

3.14.7.10   ISO 9004 clause 7.5 production and service operations

Clause 7.5.1 *Operation and realization* proposes that an organization's top management should go beyond the organization's control of realization processes in order to achieve compliance with requirements and to provide benefits to its interested parties. Examples of how this can be achieved are provided.

Clause 7.5.2 *Identification and traceability* proposes that an organization should establish an identification and traceability process to meet requirements as well as to collect data that can be used for improvement. Examples of the need for identification and traceability are provided.

Clause 7.5.3 *Customer property* proposes that an organization should identify its responsibilities relating to property and other assets that are under its control but are owned by its customers and its other interested parties. This should be done to protect the value of customer property. Examples of customer property are provided.

Clause 7.5.4 *Preservation of product* proposes that an organization's management should define and implement processes and identify resources for handling, packaging, storage, preservation, and delivery of its products in order to prevent damage, deterioration, or misuse of the products during internal processing and during final delivery. In doing so, management should consider the need for any special requirements, depending on the nature of the products. The organization should communicate to its interested parties information relating to resources and methods needed to preserve the product for its intended use throughout the life-cycle of the product.

3.14.7.11   ISO 9001 clause 7.6 control of monitoring and measuring devices

This clause covers the requirements for ensuring that equipment and other devices used for monitoring and measurement of processes and products, are capable of providing the required measurements so that valid decisions can be made with regard to acceptance of processes and products. An organization must determine the monitoring and measurement activities to be performed and the devices needed for these activities in order to provide evidence that products conform to requirements. In order to ensure that the monitoring and measurement devices give valid results, the devices must be subjected to the following:

- Calibrated or verified at specified intervals or prior to use
- Adjusted or re-adjusted as necessary
- Identified in a manner that their calibration status can be determined
- Safeguarded from adjustment that make the monitoring and measurement results invalid
- Protected from damage and deterioration during handling, maintenance and storage.

This clause is particularly applicable to food companies that operate with HACCP prerequisite programs (Chapter 4) as part of their HACCP systems.

3.14.7.12   ISO 9004 clause 7.6 control of measuring and monitoring devices

This clause proposes that an organization's management should define and implement effective and efficient measuring and monitoring processes, including the methods and devices for verification and validation of products and processes to ensure the satisfaction of its customers and its other interested parties. Examples of measuring and monitoring processes and their contents are provided.

3.14.8   ISO 9001:2000 and ISO 9004:2000 clause 8 measurement, analysis, and improvement

In *ISO 9001:2000* and *ISO 9004:2000*, clause 8 deals with monitoring and measurement, and the analysis and use of the resulting data for making decisions and for improvement on an ongoing basis.

3.14.8.1   ISO 9001 clause 8.1 general

An organization must plan, develop, and implement processes for monitoring, measurement, analysis, and improvement. These processes must be used to demonstrate that both the organization's products and its entire quality management system conform to requirements, and for continual improvement of the effectiveness of the quality management system. The organization must use the appropriate methods for monitoring, measurement, and analysis and must apply the appropriate statistical techniques for the analysis of the data.

3.14.8.2   ISO 9004 clause 8.1 general guidance

Clause 8.1.1 Introduction proposes that the top management of an organization should ensure that there is effective and efficient measurement, collection, and validation of data. These are needed to ensure the organization's performance and the satisfaction of its interested parties. The organization should continually monitor its performance improvement activities, and should use the results of data from improvement activities as information for improving its performance. Examples of performance measurement are provided.

Clause 8.1.2 *Issues to be considered* provides examples of points to be considered as part of measurement, and analysis and improvement activities.

3.14.8.3   ISO 9001 clause 8.2 monitoring and measurement

This clause requires that monitoring and measurement be applied to four areas.

Clause 8.2.1 *Customer satisfaction* specifies that an organization must monitor information relating to customer perception of whether the organization has met the customer's requirements as one of the organization's measurement of the performance of its quality management system. The organization must determine the methods for obtaining and using this information relating to customer perception.

Clause 8.2.2 *Internal audit* specifies that an organization must conduct internal audits at planned intervals to determine whether its quality management system conforms to the planned arrangements, to the requirements of the *ISO 9001* standard, and to the quality management system requirements it has established, as well as whether the system is effectively implemented and maintained. Requirements are given for the organization's internal audit program, for conducting audits, for the selection of auditors, and for actions relating to detected nonconformities and follow-up activities. A documented procedure is required for internal audit activities, and records of internal audits must be kept. This clause is particularly applicable to food companies that operate with GMPs, HACCP prerequisite programs (Chapter 4), and the HACCP system (Chapter 5).

Clause 8.2.3 *Monitoring and measurement of processes* specifies that an organization must apply methods for monitoring and, where applicable, measurement of its quality management system processes. These methods must demonstrate that the processes have the ability to achieve the results expected. Corrective action must be taken when the expected results are not achieved (see also *ISO 9001* clause *8.5.2 Corrective action*) This clause is particularly applicable to food companies which carry out process control activities as part as food processing operations, and food companies that operate with HACCP systems (Chapter 5).

Clause 8.2.4 *Monitoring and measurement of product* specifies that an organization must monitor and measure the characteristics of products at appropriate stages of the product realization processes and according to the planned arrangements, to verify that the product characteristics are met. Requirements for release of products are given. Records of product conformity and the criteria for product acceptance, and of the personnel responsible for authorizing product must be kept. This clause is particularly applicable to food companies that monitor and measure product characteristics as part of their HACCP system (Chapter 5).

3.14.8.4   *ISO 9004 clause 8.2 measurement and monitoring*

Clause 8.2.1 *Measurement and monitoring of system performance* offers guidelines in the following clauses:

- **Clause 8.2.1.1 *General*** proposes that an organization's top manage-ment should ensure that effective and efficient methods are used to identify areas for improvement of the performance of its quality management system. Examples of these methods are satisfaction sur-veys for its customers and other interested parties, internal audits, financial measurements, and self-assessment.

- **Clause *8.2.1.2 Measurement and monitoring of customer satisfaction*** proposes that an organization's management should establish effec-tive and efficient processes to collect, analyze, and use customer-re-lated information, including customer satisfaction information, for improving the performance of the organization. The organization should identify sources of customer-related information. Examples of customer-related information and of sources of customer satisfac-tion information are provided.

- **Clause *8.2.1.3 Internal audit*** proposes that the organization's top management should ensure that an effective and efficient internal audit process is established to assess the strengths and weaknesses of the organization's quality management system. The organization's management should ensure that improvement actions are taken in response to the results of internal audits. Examples of points to be considered in the internal audit process are provided.

- **Clause *8.2.1.4 Financial measures*** proposes that an organization's management should consider the conversion of data from the orga-nization's processes to financial information in order to evaluate its processes in financial terms, and to facilitate improvement of the effectiveness and efficiency of the organization.

- **Clause *8.2.1.5 Self-assessment*** proposes that an organization's top management should consider establishing and implementing self-as-sessment activities. The scope of self-assessment activities should be in relation the organization's objectives and priorities. Annex A of the *ISO 9004* standard provides *Guidelines for self-assessment.*

Clause 8.2.2 *Measurement and monitoring of processes* proposes that an organization should identify measurement methods, perform measurements to evaluate the performance of its processes, incorporate these measurements into its processes, and use the measurements in process management. The measurements of the performance of the organization's processes should be related to the needs and expectations of the organization's interested parties; examples of these measurements are provided.

Clause 8.2.3 *Measurement and monitoring of product* proposes that an organization should establish and specify the measurement requirements and the acceptance criteria for its products. The measurements of products should be planned and should be performed to verify the achievement of requirements of the organization's interested parties, and for improvement of the organization's product realization processes. Examples of points to be considered in the selection of measurement methods are provided.

Clause 8.2.4 *Measurement and monitoring of the satisfaction of interested parties.* This clause proposes that an organization should identify the measurement information required to meet the needs of its interested parties, other than its customers. Examples are provided of measurement information relating to the people in the organization, the organization's owners and investors, its suppliers and partners, and society.

3.14.8.5   *ISO 9001 clause 8.3 control of nonconforming product*

This clause establishes the requirements for an organization when it produces products that fail to meet the product requirements. A documented procedure is required by this clause, and records of the nature of nonconformities and relevant actions must be maintained. The organization must ensure that whenever any nonconforming product is detected by the monitoring and measurement activities, it is identified and controlled so that it is not used by the organization or delivered to the organization's customers until it has been dealt with by the personnel assigned the responsibility to do so. Any nonconforming product must be subjected to one or more of the following three actions:

- Elimination of the nonconformity in the product by the appropriate process followed by confirmation that the nonconformity has been eliminated
- Authorization of the use or release of the nonconforming product, or acceptance of the nonconforming product with concession by the organization's customer
- Ensuring that the nonconforming product is not used as originally intended.

If the nonconforming product has been detected after it has been used by the organization or delivered to the organization's customers, the organization must also take one or more of the three appropriate actions mentioned.

This clause is particularly applicable to food companies which operate with GMPs or with HACCP systems (Chapter 5).

3.14.8.6   *ISO 9004 clause 8.3 control of nonconformity*

This clause provides guidance for an organization to deal with situations in which there is failure to fulfill the requirements for products and processes at all stages. The guidance is provided in the following two clauses.

Clause 8.3.1 *General* proposes that an organization's top management should empower its people with the authority and responsibility to report nonconformities at any stage of a process, so that the nonconformities can be identified and addressed in a timely manner. The organization should control the identification, segregation, and disposition of nonconforming product effectively and efficiently so that these products are not misused.

Clause 8.3.2 *Nonconformity review and disposition* clause proposes that an organization's management should ensure that an effective and efficient process is established to review and dispose of identified nonconformities. The organization should ensure that the personnel who carry out review and disposition of nonconformities have the authority and resources to do so, and to define the corrective actions to address the non-conformities.

3.14.8.7   *ISO 9001 clause 8.4 analysis of data*

This clause specifies the requirements for an organization to collect and analyze data that provide information on the following:

- The satisfaction of the organization's customers
- The conformity of products to the requirements
- Process and product characteristics, trends relating to processes and products, and determination of the need for preventive action
- Performance of the organization's suppliers

Data analysis must be done on information obtained from the various monitoring and measurement activities required in clause *8.2 Monitoring and measurement*, as well as on results from other sources of measurement. The results from the analysis of data must be used to demonstrate that the organization's quality management system is effective, and identify opportunities for improvement of system effectiveness.

3.14.8.8   *ISO 9004 clause 8.4 analysis of data*

This clause provides guidelines for an organization to use in making decisions based on the results from analysis of data from measurements and monitoring, and data from other sources. Data analysis should also be used for assessing the organization's performance and to identify areas for improvement. Examples are provided.

3.14.8.9   *ISO 9001 clause 8.5 improvement*

This clause specifies the requirements for an organization to implement activities for continual improvement. The requirements are given in three sections.

Clause 8.5.1 Continual improvement specifies that an organization must continually improve the effectiveness of its quality management system through the use of its quality policy and its quality objectives, results of audits, its analysis of data, its corrective actions and preventive actions, and its management reviews.

Clause 8.5.2 Corrective action specifies that an organization must take corrective action to eliminate the causes of nonconformities, including customer complaints, in order to prevent the recurrence of the nonconformities. A documented procedure is required for corrective action activities and records of the results of corrective actions must be kept. The specific requirements for corrective action are given.

Clause 8.5.3 Preventive action specifies that an organization must determine the actions needed to eliminate the causes of potential nonconformities in order to prevent their occurrence. A documented procedure is required for preventive action activities and records of the results of preventive actions must be kept. The specific requirements for preventive action are given.

3.14.8.10   *ISO 9004 clause 8.5 Improvement*

This clause provides guidelines for an organization to implement actions in order to realize improvement in its performance on an ongoing basis. The guidelines are given in the four clauses.

Clause 8.5.1 General proposes that an organization's management should continually seek to improve the effectiveness and efficiency of the organization's processes. The organization should have a process to identify and manage its improvement activities.

Clause 8.5.2 Corrective action proposes that an organization's top management should ensure that corrective action is used as a tool for improvement. The organization should identify sources of information to define its corrective actions that should focus on eliminating the causes of nonconformities in order to prevent their recurrence; examples of these sources of information are provided.

Clause 8.5.3 Loss prevention proposes that an organization's management should mitigate the effects of loss to the organization in order to maintain the performance of its processes and products. Loss prevention activities should be planned and should be based on the various sources of data generated by the organization. Examples of sources of data are provided.

Clause 8.5.4 Continual improvement of the organization proposes that an organization's management should create a culture within the organization so that its people are actively involved in seeking opportunities for improving the performance of the organization's processes, activities, and products. The organization's top management should define and implement a process for continual improvement, and should ensure that people are empowered and have responsibility for identifying opportunities for performance improvement in the organization. Examples of inputs to support the improvement process are provided.

References

International Standard, ISO 9000, 2nd ed., 2000–12–15, Quality management systems—fundamentals and vocabulary.

International Standard, ISO 9001, 3rd ed., 2000–12–15, Quality management systems—requirements.

International Standard, ISO 9004, 2nd ed., 2000–12–15, Quality management systems—guidelines for performance improvements.

Malcolm Baldrige National Quality Award, National Institute of Standards and Technology, www.quality.nist.gov/.

chapter four

GMPs and HACCP prerequisite programs

4.1 Introduction

This chapter deals with practices commonly used to address food safety and sanitary conditions in the food industry, and particularly in food processing establishments. Many of these food safety and sanitary practices have been mandated by government food laws and regulations that prohibit the adulteration of foods. Many food safety practices are good manufacturing practices (GMPs) that have been mandated by government agencies on the basis of scientific knowledge relating to known health hazards in foods, and the need to prevent unacceptable levels of these hazards or to eliminate them from foods. In addition, some mandatory GMPs address food quality and fitness of food for human use. Some other GMPs are not mandatory but recommended practices, advisory practices, or common food industry practices; these serve as guidance for achieving food safety and food quality. In this chapter, no distinction is made between mandatory practices and recommended or advisory practices; the reader should refer to the written text of food laws and regulations for information on mandatory practices.

4.2 GMPs and government regulations

Traditionally, the food industry, and particularly the food processing sector, has relied on the use of GMPs in its efforts to ensure the safety of processed foods. Most of these GMPs are used by many national governments worldwide for monitoring the safety of consumer foods and for inspection of establishments that process, package, handle, and store foods. A good example of GMPs that are part of government regulations at the national level is the *"Current Good Manufacturing Practice In Manufacturing, Packing, Or Holding Human Food"* of the *U.S. Code of Federal Regulations (CFR), Title 21, Part 110.* The *Current Good Manufacturing Practice* can be considered the minimum criteria for the monitoring and inspection of food processing establishments by the U.S. *Food and Drug Administration* (FDA). The GMP criteria are addressed in the following *Subparts: General Provisions, Buildings and Facilities, Equipment, Production and Process Controls,* and *Defect Action Levels*; the specific topics addressed in each of these *Subparts* are listed in Table 4.1. In many situations in the food industry, GMPs are documented as Standard Operating Procedures (SOPs); for example, some food regulations in the U.S. require that certain food companies establish Sanitation standard operating procedures (SSOPs) to address some aspects of food safety.

In addition to GMPs that have been developed by national governments, food safety practices have been developed for use at the international level for the purposes of facilitating fairness in global food trade, and for the protection of health of consumers around the world. The recognized practices relating to food safety are described in the *"Recommended International Code of Practice, General Principles of Food Hygiene"* of the *Codex Alimentarius Commission, Food and Agricultural Organization/World Health Organization (FAO/WHO) Food Standards Programme.* These practices are covered in the following Sections of this standard: *Primary Production; Establishment: Design And Facilities; Control of Operation; Establishment: Maintenance And Sanitation; Establishment: Personal Hygiene; Transportation; Product Information and Consumer Awareness;* and *Training.* The specific topics addressed in each of these *Sections* are listed in Table 4.1. The GMP requirements of many national governments are compatible with the food safety practices described in this *Codex Alimentarius* standard.

4.3   GMPs as business practice in the food industry

In addition to their use for purposes of government regulations and international trade, GMPs are commonly used as part of business practices in the food industry. GMPs have been used very widely as the basis for developing and establishing food safety programs within food processing establishments. In many cases, and particularly in large- and medium-sized establishments, food manufacturers have developed food safety programs that substantially exceed the GMP requirements of governments. The objective of these programs is to meet government requirements as well as customer requirements, and to achieve competitive advantage in securing business with potential customers. Many food companies use GMP criteria as an important consideration in the selection of their suppliers of raw materials, ingredients, packaging materials, and services.

It is common practice for food manufacturers to assess and evaluate their GMP-based food safety programs for effectiveness using internal GMP audits or food safety audits. In addition, some food companies evaluate the effectiveness of the GMP-based food safety programs of their suppliers, using either second-party food safety audits or independent, third-party food safety audits. The outcome of these evaluations can have a major influence on a company's decision to purchase from a supplier or to enter into a customer-supplier relationship.

Table 4.1 Topics Covered By GMPs And HACCP Prerequisite Programs

Current Good Manufacturing Practices in Manufacturing, Packing, Or Holding Human Food, U.S. Code of Federal Regulations, Title 21 CFR Part 110

Subpart A-General Provisions
– Definitions
– Current good manufacturing practice
– Personnel
– Exclusions

Subpart B-Buildings and Facilities
– Plants and grounds
– Sanitary operations
– Sanitary facilities and controls

Subpart C-Equipment
– Equipment and utensils

Subpart E-Production and Process Controls
– Process and controls
– Warehousing and distribution

Subpart G-Defect Action Levels
– Natural or unavoidable defects in food for human use that present no health hazard

Examples of Common Prerequisite Programs for HACCP systems, U.S. National Advisory Committee on Microbiological Criteria for Foods (NACMCF, 1997)

– Facilities
– Supplier control
– Specifications
– Production Equipment
– Cleaning and Sanitation
– Personal hygiene
– Training
– Chemical control

– Receiving, storage, and shipping
– Traceability and recall
– Pest control
– Others: quality assurance procedures; labeling; standard operating procedures for sanitation, processes, product formulations, and recipes; glass control; employee food and ingredient-handling practices

Recommended International Code of Practice, General Principles of Food Hygiene, Codex Alimentarius Commission (1997)

Establishment: Design and Facilities
– Location
– Premises and rooms
– Equipment
– Facilities

Control of Operations
– Control of food hazards
– Key aspects of hygiene control system
– Incoming materials requirements
– Packaging
– Water
– Management and supervision
– Documentation and records
– Recall procedures
– Incoming materials requirements

Establishment: Personal Hygiene
– Health status
– Illness and injuries
– Personal cleanliness
– Personal behavior
– Visitors

Transportation
– General
– Requirements
– Use and maintenance

Product Information and Consumer Awareness
– Lot identification
– Product information
– Labelling
– Consumer education

-- continued

Table 4.1 Topics Covered By GMPs And HACCP Prerequisite Programs

Current Good Manufacturing Practices in Manufacturing, Packing, Or Holding Human Food, U.S. Code of Federal Regulations, Title 21 CFR Part 110

Establishment: Maintenance and Sanitation

– Maintenance and cleaning	*Training*
– Cleaning programs	– Awareness and responsibilities
– Pest control systems	– Training programs
– Waste management	– Instruction and supervision
– Monitoring effectiveness	– Refresher training

Sources: *J. Food Prot.* 61, 9, 1998, 1246–1259; and the Food and Agriculture Organization of the United Nations, codex Alimentarius, *1997 Basic Texts on Food Hygiene,* Codex Alimentarius Commission, Join FAO/WHO Food Standards Program, Rome.

4.4 GMPs and HACCP prerequisite programs

With the use of the hazard analysis critical control point (HACCP) system to address food safety in food processing, GMPs have become part of the very basic requirements that must be in place before an effective HACCP system can be implemented. Consequently, the traditional GMPs, along with some additional requirements, are now universally regarded as *Prerequisite Programs* for the implementation of an HACCP system. In the U.S., the mandatory HACCP systems for certain food industry sectors have SSOPs as part of the prerequisite programs.

The emergence of the HACCP system for achieving food safety has reinforced the importance of GMPs to the extent that the prerequisite programs are considered as the necessary foundation on which an effective HACCP system is built. In a functional HACCP system at a food establishment, the GMPs are part of the HACCP prerequisite programs that address all food safety considerations that are not included as a part of the HACCP plan(s), as well as all unsanitary and undesirable but nonhazardous contaminants. Together, the HACCP prerequisite programs and the HACCP plan(s) at a food establishment form the *"House of Product Safety"* (*ASQ Food, Drug, and Cosmetic Division, 2002*).

4.5 Activities covered by GMPs and HACCP prerequisite programs

When GMP programs or HACCP prerequisite programs are developed and implemented at a food plant, they should cover the basic controls required for raw materials, ingredients, packaging materials and products, and for the plant's facilities, employees, equipment, operations, and environment that influence the safety of a food. Many of the GMPs and prerequisite program activities are directed at ensuring that the necessary conditions exist for the prevention of potential contamination and cross-contamination of food.

The activities addressed by HACCP prerequisite programs have been well documented by various organizations, food companies, and government regulatory agencies; GMPs are a substantial part of these prerequisite programs. However, a standardized framework for HACCP prerequisite programs has not been developed in the same way as the formal HACCP system, with its recognized set of seven principles. Consequently, the actual framework for designing HACCP prerequisite programs tends to vary from one country to another and from one food industry sector to another. However, of greater importance is the fact that the topics the prerequisite programs cover and the activities they address are essentially very similar. This point is best illustrated by the following examples:

- In the *U.S. Department of Agriculture (USDA)/Food Safety and Inspection Services (FSIS)* mandatory *Pathogen Reduction: HACCP* inspection program for the meat and poultry sector, HACCP prerequisite programs are covered in sections on *Establishment's grounds and facilities, Equipment and utensils, Sanitary operations, Employee hygiene, Tagging unsanitory equipment, utensils, rooms or compartments* and mandatory *Sanitation standard operating procedures (SSOPs)*.

- In the *FDA's Seafood HACCP Final Rule*, HACCP prerequisite programs are addressed by mandatory *Current good manufacturing practice* (Table 4.1) and *Sanitation control procedures*, consisting of recommended *Sanitation SOP* and mandatory *Sanitation monitoring*.

- In the *FDA's Juice HACCP Final Rule*, HACCP prerequisite programs are addressed by mandatory *Current good manufacturing practice* (Table 4.1) and mandatory *Sanitation standard operating procedures*.

- The *NACMCF's Hazard Analysis and Critical Control Point Principles and Application Guidelines in 1997*, recognized the following as *Examples of Common Prerequisite Programs: Facilities, Supplier Control, Specifications, Production Equipment, Cleaning and Sanitation, Personal Hygiene, Training, Chemical Control, Receiving, Storage, and Shipping, Traceability and Recall, Pest Control*. (Table 4.1)

- In Canada, *Agriculture and Agri-Foods Canada Food Safety Enhancement Program (FSEP)*, HACCP prerequisite programs are covered by the following: *Premises, Transportation and Storage, Sanitation, Equipment, Personnel Training, Recall Program* and *Records*

- At the international level, the *Codex Alimentarius Commission's Recommended International Code of Practice General Principles of Food Hygiene* can be considered as the HACCP prerequisite programs.

For purposes of comparison, Table 4.1 lists the activities that are recognized as *Current good manufacturing practice (Title 21 CFR Part 110)* and as HACCP prerequisite programs, including the activities covered by the *Codex Alimentarius Commission's General Principles of Food Hygiene*.

4.6 GMPs, HACCP prerequisite programs, HACCP systems, and quality systems

Food plants that operate with HACCP systems will have the required prerequisite programs that include GMPs and in some industry sectors in the U.S., also include SSOPs. In food processing plants that do not have HACCP systems, the GMPs remain the essential practices for addressing food safety. In these situations, it is more appropriate for the HACCP prerequisite programs to be used instead of the basic GMPs. The prerequisite programs provide more comprehensive coverage of the requirements relating to conditions and activities in a food plant than will the GMPs. In food plants that do not operate with HACCP systems but have implemented the *ISO 9001 Quality management systems* standard, as a minimum the GMPs should be part of the applicable regulatory requirements of the standard.

Within the requirements of the *ISO 9001:2000* standard, there are specific sections that address activities that can be related to activities covered by some of the HACCP prerequisite programs. This is evidence of compatibility between the quality management system requirements and HACCP prerequisite programs. In addition to the use of the *ISO 9001* standard by some food companies, many other food companies have developed their own companywide quality systems that include the activities covered in the HACCP prerequisite programs.

From the discussion in the preceding paragraph, and on the basis of the food safety and quality activities that are used in their operations, food plants can be identified within one of the following categories:

- Food plants which operate only with the mandatory GMPs required by government regulations, and other food safety and quality practices required by customers.
- Food plants that operate with HACCP systems required by government regulations or by customers, and quality practices required by government and customers. The HACCP prerequisite programs include GMPs.
- Food plants that operate with voluntary HACCP systems and quality practices required by government and customers. The HACCP prerequisite programs include the GMPs
- Food plants that operate with quality systems such as *ISO 9001:2000* or other nonregistered quality systems which address quality practices required by government and customers. As a minimum, GMPs should be integrated into the quality systems.
- Food plants which operate with quality systems and mandatory or voluntary HACCP systems. The HACCP prerequisite programs include the GMPs.

4.7 Development and implementation of GMPs and HACCP prerequisite programs

All food plants need to develop and implement a program of GMPs to address food safety requirements. Food plants that develop and implement HACCP systems to address food safety also need to develop and implement the HACCP prerequisite programs. In the U.S., food plants identified within some food industry sectors, such as the meat and poultry sector, the juice processing sector, and the seafood sector, are required to operate with HACCP systems that include specific mandatory prerequisite programs that contain SSOPs. It is the responsibility of a food plant's senior management to develop, implement and maintain the appropriate food safety program, whether it is a GMP-based program or a complete HACCP system. The food plant's employees who are responsible for the food safety program must be familiar with the government regulatory requirements that apply to the operations at the plant.

The activities for GMPs or the HACCP prerequisite programs that are developed at a food plant should be documented as SOPs. In addition, where monitoring, inspection, or testing is carried out as part of the programs, records should be kept as evidence that these activities are actually performed. The documents and records that are used in these programs should be controlled in the same manner as documents and records are controlled in a quality system. The prerequisite programs should be developed as an integrated set of activities that address food safety requirements not specifically addressed by HACCP plans (Chapter 5).

In general, many of the activities which are included in GMPs program or in the HACCP prerequisite programs should be quite similar for most food plants, regardless of the type of processing that is carried out. Therefore, it is possible to compile a generic list of topics which address the entire range of activities in a particular prerequisite program or a particular category of GMPs. However, certain specific activities in this generic list might not be applicable to certain food plants; this depends on the specific type of processing that takes place in a particular food plant. In addition, in certain situations it is possible that a particular activity in a prerequisite program can be considered as a critical control point (CCP, Chapter 5) in an HACCP plan for a product at a particular food plant.

The following sections describe the activities of a set of generic HACCP prerequisite programs; they include the prerequisite programs recognized by the NACMCF (1997), and also address the activities covered in the various sections of the *Codex Alimentarius General Principles of Food Hygiene* standard. In addition, the prerequisite programs that are described in the following sections include the GMP activities required by food regulations, and should, therefore, be applicable to food plants that operate without HACCP systems; consequently, these sections are titled as HACCP prerequisite programs.

4.8 HACCP prerequisite program premises and facilities

This prerequisite program addresses the requirements for the location, design, construction, and maintenance of buildings that are used for food processing. It covers the grounds, all exterior and interior structures of buildings, and all facilities and essential services required in food processing establishments. Many sections of this prerequisite program need to be considered during the design and construction stages of a food plant. Engineering, design, and construction requirements and guidelines for food plants have been developed by some government agencies.

The general conditions in a food plant, including the state of repairs, maintenance, and cleanliness of all structures and facilities, are critical in order to achieve the sanitary requirements for food processing. The primary considerations are that the building, its grounds, structures, and facilities are not a source of contamination or cross-contamination of food, there is protection from entry of pests into the building, and clean and sanitary conditions can be maintained.

This prerequisite program includes the GMP items addressed in the sections on *Plants and grounds* and *Sanitary facilities and controls* of *Subpart B-Buildings and Facilities* of the *Current Good Manufacturing Practices* (Table 4.1), and in *Establishment: Design and Facilities* of the *General Principles of Food Hygiene* (Table 4.1). Table 4.2 provides a list of the topics which are covered by this prerequisite program.

Relationship to ISO 9001:2000: The requirements of *6.3 Infrastructure* address the topics covered by this prerequisite program.

4.8.1 Location

The buildings in which food is processed or stored should not be located in close proximity to sites that are sources of environmental pollutants, pest infestations, smoke or dust, to areas that accumulate wastes or stagnant water, or have industrial, agricultural or other activities which are potential sources of food contamination. If any of these undesirable conditions exist, there should be adequate safeguard to protect against any potential contamination or pest infestation.

4.8.2 Grounds

The grounds of buildings in which food is processed or stored should be adequately sloped and drained to prevent stagnant water, be free of waste and debris, be controlled for dust, and be adequately maintained to protect against becoming a source of contamination or pest infestation. In order to protect from pests, maintenance of the grounds should address grass and lawns, hedges, shrubs, trees, receptacles for storage of garbage, and any structures located on the grounds. In particular, the perimeter of the building exterior should be well maintained to prevent breeding or attraction of pests.

Table 4.2 Topics Covered by Prerequisite Program
Premises and Facilities

Location
Grounds
Building exterior
Building interior
– Design and layout
– Structures
– Glass
– Corners and joints
– Floors
– Windows and doors
– Lighting
– Ventilation
– Drains
– Pipes and hoses
Access to premises
Employee facilities
– Hand-washing stations
– Washrooms and change-rooms
– Lunchrooms and break-rooms
Cleaning and sanitizing facilities
Storage facilities
Waste collection/storage facilities
Water/steam/ice
– Supply
– Quality

Parking lots and roadways should be maintained so that they are not sources of dust and airborne contaminants that can enter the building.

4.8.3 Building exterior

The design, construction and maintenance of exterior walls and roofs of buildings, should prevent the entry of sources of contamination and pests, and leakage of water into the building. Exterior walls should be free of cracks that could be breeding sites for pests. Openings for exhaust fans and air intake ducts, and exterior drainpipes should be adequately screened and protected to prevent entry of pests. The loading and unloading areas of the building, and all exterior doors and windows should be adequately protected to prevent entry of pests. The location of exterior lighting should not attract insects into the building.

4.8.4 Building interior

- *Design and layout*: The design and layout of the building interior and the location of all structures, equipment, services, and facilities should permit movement of personnel and equipment, flow of air, materials and products, movement of waste and garbage, and storage of materials and products, in a manner that prevents contamination and cross-contamination of materials and products. There should be designated areas for processing, for packaging, for raw materials storage and preparation, and for finished product storage. Processing activities that are potential sources of cross-contamination should be located in areas that are separated from other processing activities. The layout of the building should provide adequate workspace for plant employees to perform their tasks satisfactorily, and adequate space for cleaning of all structures. The location of equipment should be such that it is accessible for regular cleaning and maintenance.

- *Structures*: The design and construction of all building structures such as floors, walls, ceilings, overhead structures, windows, doors, and stairs, and all utilities and service structures such as ducts, pipes and drains should meet all requirements for construction of food plants. The materials used for construction and finishing of these structures should not be sources of contamination and should be durable, impervious, smooth and easily cleaned and maintained. The materials used for construction of walls and floors should withstand the routine plant operating conditions and the routine cleaning and sanitizing conditions.

- *Glass*: Glass or glass-like materials such as breakable plastic in food plants can be sources of physical hazards and should not be used in processing areas where there is a likelihood of breakage that will result in contamination of product. If these materials must be used, they must be adequately protected from breakage. Overhead light bulbs should be protected from breakage if they are a potential source of contamination. The control of glass and glass-like materials in a food plant should be addressed in a glass policy that should cover instructions for dealing with breakage of glass, and the control of glass and glass-like materials in items such as clocks, lights, gauges, containers and glassware.

- *Corners and joints*: Corners and joints in all structures should be designed to prevent accumulation of contaminants and to facilitate cleaning; they should be free of cracks and openings. The junctions between walls and floors should be designed to facilitate cleaning. Joints on walls, floors, and ceilings should be sealed and should be easily cleaned.

- *Floors:* The surface of floors should be even, but with the appropriate slope for waste-water and other liquids to be drained at the designated outlets. Floor surfaces should be impervious, durable and free of cracks to facilitate cleaning.

- *Windows and doors:* Windows on exterior walls should be sealed or fitted with screens to prevent entry of pests. Exterior doors should be self-closing, should always be kept closed, and should be without gaps or openings when closed, to prevent entry of pests. Exterior doors should be kept closed to prevent unauthorized access into the building.

- *Lighting:* The building interior should be equipped with adequate light and lighting facilities to permit employees to carry out their designated tasks in areas where processing, handling, storage, testing, inspection and cleaning activities take place. Adequate lighting should also be provided in hand-washing areas, change-rooms, locker rooms and toilet rooms.

- *Ventilation:* There should be adequate ventilation and air exchange throughout the building to prevent airborne contamination, condensation on any structure or equipment, and accumulation of dust. High humidity should be avoided to prevent mold growth and some types of insects. The direction of air-flow should not result in contamination or cross-contamination of foods. In addition, fans and other air-blowing equipment should be operated in a manner that does not result in contamination. Air filters and dust collectors should be cleaned, maintained and replaced so that they are effective and do not become sources of contamination.

- *Drainage and sewage systems:* The drainage and sewage systems should be designed to prevent cross-connection of sewage with other wastes from the plant in order to avoid any potential for contamination. Drains should be adequately sloped to ensure there is no accumulation of wastewater or other liquids. The location of drains and drain traps should permit ready access for cleaning. The design and maintenance of the drainage system should prevent backflow of wastewater into the building.

- *Pipes and hoses:* Water pipes should be free of condensation. Insulated pipes should be well maintained and should be free of condensation drips and mold growth. Water taps, faucets, and hose connections should be free of leaks and water drips. Hose reels should be provided for storage of water hoses when not in use.

4.8.5   Access to premises

The entrances and exits of a food plant should be controlled to prevent access by unauthorized personnel. Exterior doors should not open from the outside of the building. Food plant employees should use only the designated entrances and exits. Receiving and shipping locations should not be used by employees as entrances or exits. Access of visitors into food plants should be controlled.

4.8.6   Employee facilities

- *Hand-washing*: There should be accessible hand-washing stations at the appropriate locations, with potable running water at a suitable temperature, soap or other hand-cleaning and sanitizing materials, sanitary hand-drying equipment or supplies for employees to wash and dry hands as required. The water control devices at the hand-washing stations should be designed to protect against recontamination of washed hands. If disposable towel is used for hand-drying, a covered garbage receptacle should be provided for used towels. There should be easily understandable signs posted at hand-washing stations to remind employees to wash hands.

- *Washrooms and toilet rooms*: Washrooms and toilet rooms in particular, should be separated from and should not open directly into food storage, handling and processing areas. Washrooms should be equipped with the required hand-washing facilities, covered garbage receptacles, and easily understandable signs to serve as reminders to employees.

- *Change-rooms: Change*-rooms should be available for employees to change from their personal external clothing into designated work uniform and footwear. Change-rooms should be equipped with lockers or suitable storage racks for employees to store clothing, footwear and other personal items, with receptacles for dirty work clothes, and covered garbage receptacles. The design and location of lockers and storage racks in change-rooms should facilitate cleaning.

- *Lunchrooms and break-rooms*: There should be designated lunchrooms and break-rooms for employees. Lunchrooms should be equipped with appropriate appliances and food storage facilities for employees' food and with covered garbage receptacles. If smoking is permitted in lunchrooms and break-rooms, it should be restricted to designated areas and ashtrays should be provided.

4.8.7   Cleaning and sanitizing facilities

Potable running water at the required temperatures and pressures should be available for all cleaning and sanitizing activities. The required equipment and tools for cleaning and sanitizing should be available. Equipment and tools used for cleaning of food-contact surfaces, food processing equipment and utensils should be appropriately identified and stored so that they are separate from those used for cleaning of building structures such as floors and walls. There should be designated areas for cleaning of cleaning equipment and tools and for their storage when they are not in use.

4.8.8   Storage facilities

There should be adequate and appropriate facilities such as warehouse, storage rooms, silos, tanks, vats, bins, or other containers, for the storage of raw materials, ingredients, packaging materials, products to be reworked or recycled, semi-finished products, finished products, cleaning materials and nonfood chemicals. These storage facilities should be designed to ensure that there is no contamination, cross-contamination, or pest infestation of raw materials, ingredients, packaging materials, and semi-finished and finished products during storage. There should be separate storage facilities to segregate food materials from nonfood chemicals.

There should also be appropriate storage facilities for idle food processing equipment and for tools, materials, and spare parts used for repair and maintenance of equipment.

4.8.9   Waste collection and storage facilities

There should be designated containers with covers, if necessary, for collection of waste and garbage and for their temporary storage until disposal. These containers should be properly identified, and be made of durable, impervious material and maintained in a sanitary condition. There should be no leakage from waste containers. Waste collection containers located on the grounds outside of the building should be maintained so that they are not sources of contamination or pest infestation.

4.8.10   Water

- *Water supply*: There should be an adequate supply of potable water, at the desired temperatures and pressures, for use in processing operations and for cleaning. There should be facilities to ensure that temperature and pressure requirements for water can be achieved.

- *Water quality*: Only potable water should be used in all food plant processing and cleaning operations. The water quality should conform to the guidelines for potable water based on microbiological, chemical, and physical specifications of applicable government agencies. Water should be tested periodically to determine if it complies with these specifications; the records of water quality test results should be maintained.

If water treatment facilities are located on the premises, they should be adequately monitored and maintained; records should be kept for water treatment activities. Water treatment processes that are designed for re-circulated water should have the required controls with the appropriate records to demonstrate compliance to the re-circulated water quality specifications.

Only approved chemicals should be used for treatment of water used in food plants. Containers with chemicals for water treatment should be identified with labels, and their conditions and location of storage should prevent any potential contamination of food or food contact surfaces.

4.8.11   Ice and steam

Ice for use in food plants should be made from potable water and should be handled and stored to protect from contamination. Steam that comes into contact with food or food contact surfaces should be generated from potable water. Only approved chemicals should be used in boilers which generate steam for these purposes; these chemicals should be controlled in a similar manner as described for water treatment chemicals.

4.9   HACCP prerequisite program personnel training, hygiene and practices

This prerequisite program addresses the requirements for employees of food plants; some of the requirements are also applicable to visitors to food plants as well as any other personnel who are not employees at a food plant, but carry out some type of work on the premises or facilities. The requirements for personnel at food plants can be conveniently classified under the sections, *Personnel training*, *Personal hygiene*, and *Personnel practices*.

Employees in a food plant play a critical role in ensuring the safety of foods produced at the plant. In addition, employees should not contribute to or be a source of contamination or cross-contamination of foods. In this prerequisite program, the primary considerations are to ensure that both temporary and permanent employees have the required education and training, are adequately supervised, and follow their required work-related tasks, personal hygiene requirements and acceptable personnel practices during their work.

This prerequisite program covers the items included in the section on personnel in *Subpart A—General Provisions of the Current Good Manufacturing Practice* (Table 4.1) and in *Section VII—Establishment: Personal Hygiene* and *Section X—Training of the General Principles of Food Hygiene* (Table 4.1). Table 4.3 lists the topics covered by this prerequisite program.

Relationship to ISO 9001: The requirements of *6.2 Human resources* address the topics covered by this prerequisite program.

4.9.1   Personnel training

- *Food safety training*: All food plant employees, including temporary employees, should be trained in the basic food safety principles and practices that are required to prevent contamination and cross-contamination of foods. This training should cover hygienic food handling practices, personal hygiene requirements, and the dangers

Table 4.3 Topics Covered by Prerequisite Program
Personnel Training, Hygiene, Practices

Personnel training
– Food safety training
– Technical training
– Training records
Personal practices
– Personal hygiene
– Hand washing
– Eating and smoking
– Garment and work-wear
– Personal items
Illness and injuries
Visitors and noncompany personnel
– Controlled access to premises
– Personal practices

associated with poor personal hygiene and unsanitary personnel practices in a food plant. In addition, personnel who have the responsibility to monitor the adequacy of food safety practices of food plants should have the necessary training and experience to recognize and identify food hazards and situations that have the potential to lead to contamination or cross-contamination of foods. This includes the training of supervisory personnel to recognize injuries or infectious illnesses among plant employees. The food safety training needs of employees in food plants should be reviewed periodically, and if necessary, additional training or refresher training should be provided. Whenever food safety training is provided to employees, there should be an evaluation to determine that the training is understood and could be put into practice by the employees.

- *Technical training*: Employees whose tasks involve operation, maintenance, and cleaning of food processing equipment, and sanitation and cleaning activities should be provided with the relevant technical training that is required to carry out their specific tasks so that all food safety requirements are met. The technical training needs of employees should be reviewed periodically, and if necessary, additional or refresher technical training should be provided. Technical training should also include on-the-job training and evaluation to ensure that employees understand the training and perform their tasks based on the training.

- *Training records:* Records should be kept as evidence that the relevant food safety training or technical training was provided to employees, and that they were evaluated after completion of the training. In addition, there should be records to show that the training needs of employees are reviewed periodically.

4.9.2 Personnel practices

- *Personal hygiene*: In order to protect against contamination of products, food plant employees are required to maintain satisfactory personal grooming and cleanliness and to practice good personal hygiene habits during all food handling operations. This includes general cleanliness of clothing and body, including hair and fingernails. Employees should refrain from placing fingers in mouth, nose or ears, and from eating, chewing, spitting, and smoking during food handling operations, and avoid coughing and sneezing over unprotected products, food-contact surfaces, or food processing equipment.

- *Hand-washing*: In order to protect hands from being a source of contamination of products, food plant employees should wash, sanitize if necessary, and dry their hands at the designated hand-washing stations when their hands become dirty. Employees should wash hands before start of work, when re-entering their work area, after a visit to the toilet, after coughing or sneezing into their hands, or after handling raw materials, equipment, waste or waste containers, or after any other situation that will cause the hands to become dirty and be a source of contamination or cross-contamination.

- *Eating, drinking and smoking*: Employees should eat, drink, and if permitted, smoke only in the designated lunchrooms and break-rooms or other authorized areas. Employees' food or drink should be kept in the designated areas and should not be taken into their work areas. Drinking of water should be done at the designated water fountains. Employees should not take their medication into their work areas.

- *Garments and work-wear:* Employees should wear the uniforms or outer garments provided for their work. Uniforms and garments should be clean at the start of work and should be changed when they become dirty or according to the required change frequency. If gloves are required to be worn during work, they should always be clean and sanitary, and should be changed if they become torn. Hair and beard restraints should be worn to completely cover hair and beard.

- *Personal items:* Employees should not wear jewelry, hairpins, wristwatches or other personal items such as false eyelashes, false fingernails and nail polish during food handling operations. In addition, during work, employees should refrain from keeping in their possession any personal items which could be a potential source of contamination.

4.9.3 Illness and injuries

Food plant employees with certain illnesses or injuries should be excluded from food handling activities; these illnesses include jaundice, diarrhea, vomiting, fever, sore throat with fever, open or infected skin lesions (e.g., boils, cuts, burns), discharges from the eyes, ears or nose, or any disease that can be transmitted through food. Employees should inform their supervisors if they suffer from any of these health conditions. In addition, supervisory personnel in food plants should constantly monitor food-handling employees for these injuries and illnesses.

4.9.4 Visitors and noncompany personnel

- *Controlled access to premises:* The access of visitors and noncompany personnel to a food plant should be controlled to avoid any potential source of contamination from these individuals. This control should apply to family members of employees, noncompany personnel working on the premises, suppliers, customers, government inspectors, auditors, and individuals from educational and other institutions on organized visits and plant tours.

- *Personal practices*: Visitors and noncompany personnel who are permitted entry into the food processing and handling areas of the food plant should be required to follow the same personal practices of the regular employees of a food plant. These include practices relating to personal hygiene, hand-washing, eating, drinking, smoking, outer garments, personal items and illnesses and injuries.

4.10 HACCP prerequisite program sanitation and cleaning

This prerequisite program covers all ongoing and periodic activities and operations that are directed at maintaining the environment, facilities, structures, and equipment in a food plant under sanitary conditions at all times. The design, construction, and layout considerations that relate to sanitation and cleaning are covered in the prerequisite program *Premises and Facilities*. It is quite common for the term sanitation alone to include house-keeping, cleaning and sanitizing; however, the distinction should be made between cleaning and sanitizing. In general, cleaning activities cover the removal of dust, dirt, debris, accumulated raw materials, ingredients or product, and any chemical residues, from utensils, food processing equipment, and structures. Sanitizing activities cover the use of a chemical agent or a specific technique to kill microorganisms present on equipment, utensils and structures.

The maintenance of sanitary conditions in a food plant, including clean and sanitary environment, structures, facilities, and equipment is essential to ensure that food is produced under sanitary conditions, to prevent contamination from these sources and to prevent breeding of pests. Food processing operations in a food plant should only commence after all the required cleaning and sanitizing activities have been completed. In this prerequisite program, the primary considerations are the activities for maintaining sanitary conditions by means of a sanitation program, and ongoing monitoring of the sanitary conditions during all operations at a food plant.

Sanitation standard operation procedures (SSOPs), a prerequisite program, covers some of the items included in the section on *Sanitary operations* in the *Current Good Manufacturing Practices* (Table 4.1) and in *Establishment: Maintenance and Sanitation* of the *General Principles of Food Hygiene* (Table 4.1). Table 4.4 lists the topics that are covered under this prerequisite program.

Relationship to ISO 9001: The requirements of *6.4 Work environment* address the topics covered by this prerequisite program.

4.10.1   Sanitation and cleaning program

There should be a written program for cleaning and sanitizing of the structures, facilities, and equipment in a food plant. This written program should identify each structure, facility, and equipment to be cleaned and sanitized. For each of these, the program should include detailed cleaning and sanitizing procedures, the cleaning and sanitizing chemicals to be used and their concentrations or dilutions, removal of residues of cleaning and sanitizing chemicals, the cleaning tools to be used, the frequency of cleaning and sanitizing, and the personnel responsible for cleaning specific equipment or structures. Each aspect of the cleaning and sanitation program should be monitored to ensure that the program is followed. There should be some verification to determine the effectiveness of the program. Sanitation records should be kept for the activities that are required, including the monitoring and verification of the sanitation program.

Table 4.4 Topics Covered by Prerequisite Program
Sanitation and Cleaning

Sanitation and cleaning program
Equipment cleaning and sanitizing
– Cleaned-in-place (CIP) equipment
– Cleaned-out-of-place (COP) equipment
– Utensils and food contact surfaces
Master cleaning schedule
– Daily housekeeping
– Establishment cleaning
Cleaning and sanitizing chemicals
Cleaning tools and equipment
Cleaning area
Cleaning and sanitizing personnel
Effectiveness of cleaning and sanitizing
Sanitation and cleaning records

4.10.2 Equipment cleaning and sanitizing

There should be a program for cleaning of all equipment, including all food processing and food handling equipment, as well as food storage equipment such as storage tanks, refrigerators, and freezers. For each type of equipment there should be a schedule for cleaning and sanitizing, along with any specific cleaning instructions.

- *Cleaned-in-Place (CIP) Equipment*: In some food processing operations, certain equipment cannot be easily disassembled for cleaning between successive production runs, although this equipment should be cleaned at the end of each use. These types of equipment are considered CIP equipment. There should be written procedures for cleaning and sanitizing of each type of CIP equipment.

- *Cleaned-out-of-Place (COP) Equipment*: Certain food processing equipment should be disassembled for cleaning after each use or periodically, these are considered COP equipment. There should be instructions for disassembly and reassembly of COP equipment, in addition to instructions for their cleaning and sanitizing.

- *Utensils and food contact services*: All utensils used for handling food, including all containers, trays, pans and dollies, and all food contact surfaces should be cleaned and sanitized as it becomes necessary. These should be protected from contamination if they are stored after cleaning.

4.10.3 Master cleaning schedule

- *Daily housekeeping*: A program of daily cleaning should be in place to ensure that equipment and work areas are maintained in a clean state during routine daily operations so as to prevent contamination of products. This may require that equipment, utensils, food contact services, work area and floors be cleaned periodically during daily operations. Clean up of leaks and spills of all types of materials and products should be done as soon as possible, or immediately if they are potential sources of contamination or pest infestation. There should be monitoring and verification of this daily housekeeping program.

- *Establishment cleaning*: In addition to the daily housekeeping, a comprehensive program should be in place to ensure that every aspect of the establishment is subject to periodic cleaning to eliminate the potential for contamination. This program should cover all facility structures, including lunch rooms, appliances, vending machines, break-rooms, change-rooms, and toilet rooms, ceilings and overhead structures such as pipes and ducts, walls and floors of all structures, windows and doors, and all equipment. The frequency and type of cleaning should be specified for each facility or structure. There should be monitoring and verification of this cleaning program for the establishment.

4.10.4   Cleaning and sanitizing chemicals

All chemical compounds used for cleaning and sanitizing should be approved as safe for use in food establishments, and on food-contact surfaces in particular, by an appropriate government regulatory agency. The manufacturer's guidelines and directions for use of these chemicals must be followed to ensure the effectiveness of the cleaning and sanitizing, and to remove or prevent potential contamination. The chemicals themselves must not be a source of contamination.

All packages and containers, including intermediate containers with chemical compounds used for cleaning and sanitizing, must be clearly identified by labeling, and must be stored separately from food materials and products.

4.10.5   Cleaning tools and equipment

All tools and equipment (e.g., brushes, dustpans, brooms, mops, trays, carts) used for cleaning should also be subjected to appropriate cleaning and storage. Broken or damaged tools should not be used. Tools used for cleaning food processing, storage and handling equipment should not be used for other cleaning, should not be stored on the floor, and should be identified and stored separately from tools used for other cleaning. In addition, tools used for cleaning of toilets and toilet rooms should be identified and not be used for cleaning of food processing, handling and storage areas.

4.10.6   Cleaning area

There should be designated location and facility for cleaning of cleaning tools and equipment. This area or facility, as well as all sinks and washbasins and surrounding areas, should be kept clean and sanitary. Wash-water should be removed and drained immediately to prevent the potential for contamination.

4.10.7   Cleaning and sanitizing personnel

Personnel who are assigned cleaning and sanitizing tasks should be trained in the safe use of the cleaning and sanitizing chemicals and the proper handling, identification and storage of these chemicals. They should be provided with the directions for use, including the appropriate usage concentration or dilution for food-contact surfaces, and instructions for removal of residues of these chemicals from these surfaces.

4.10.8   Effectiveness of cleaning and sanitizing

The effectiveness of the cleaning and sanitizing activities for removal of contamination should be verified. This should be done by microbiological swab tests, by visual inspection of cleaned equipment and areas, and by observing employees who carry out the cleaning and sanitizing activities.

In addition, there should be no unacceptable residues of cleaning and sanitizing chemicals on food-contact surfaces.

4.10.9   Sanitation and cleaning records

As part of the establishment's cleaning and sanitation program, records should be kept as evidence that the activities in the program are performed according to the required instructions and frequency, and that the program is monitored routinely and verified for its effectiveness.

4.11 HACCP prerequisite program pest control

This prerequisite program covers the specific activities that are directed at controlling, preventing and excluding the occurrence of pests, particularly rodents, insects and birds, from a food plant. Pet animals such as cats and dogs should not be allowed to enter food plants. The prerequisite program *Premises and facilities* includes certain preventive measures for pest control; these relate to the building structures, as well as the internal environment and the external surroundings. Pest control measures are also part of the prerequisite programs *Sanitation and cleaning* and *Transportation, receiving, storage, and shipping*.

The pest control program includes the specific activities directed at detecting pests and pest activity, both within a food plant and its immediate exterior, preventing pests from entering the building, eliminating pests from the building and the immediate surroundings, and monitoring of the pest control program for its effectiveness. In addition, the program includes control of the use and storage of chemicals or other materials used for pest control to prevent contamination of product.

This prerequisite program is included in the section on *Sanitary operations* in *Subpart B—Buildings and Facilities* in the *Current Good Manufacturing Practices* (Table 4.1) and in *Pest control systems* in *Establishment: Maintenance and Sanitation* of the *General Principles of Food Hygiene* (Table 4.1). Table 4.5 lists the topics which are covered under the prerequisite program Pest Control.

Relationship to ISO 9001: The requirements of *6.4 Work environment* address the topics covered by this prerequisite program.

4.11.1   Pest control program

There should be a formal, documented pest control program that is maintained for the establishment. This program should cover all of the preventive measures that are taken to exclude and eliminate pests, the various pest-control devices and pest-control chemicals that are used, the monitoring of pest activity, and compliance with government regulations on use of pesticides and pest control devices. It is common for food companies to engage the services of external pest control contractors to undertake some of the required pest-control activities.

Table 4.5 Topics Covered by Prerequisite Program
Pest Control

Pest control program
– Pest control devices
– Monitoring and maintenance of devices
Pest control personnel
Pest control chemicals
Monitoring for effectiveness
Pest control records

- *Pest-control devices*: The establishment's pest control program should include outside bait stations for rodent control, nettings, bait stations or mechanical traps for birds, inside devices such as mechanical traps, glue-boards for rodents, and insect light traps for flying insects. These devices should be located at appropriate positions where they are most effective for removing pests from the building. Outside devices should be located so as to prevent entry of pests into the building. There should be an updated diagram or map to show the actual locations of all pest control devices both inside and outside the building.

- *Monitoring and maintenance of devices*: The pest control devices should be monitored at an established frequency for any pest activity and for the status of the devices. If this monitoring shows unusual pest activity, the appropriate follow-up action should be undertaken immediately. In addition, if the pest control device has lost its effectiveness, it should be serviced. A written report of this monitoring and maintenance should be kept.

4.11.2   Pest control personnel

The personnel responsible for placement, monitoring and maintenance of the pest control devices, and for handling and use of pesticides, should have the required qualification and training, including food safety training. If an external pest control company is contracted by a food plant to carry out the activities of the pest control program, it should be required to provide evidence that it has the appropriate license or certification from the appropriate government agency.

4.11.3   Pest control chemicals

Only chemicals approved as pesticides by the appropriate regulatory agency should be used for pest control in a food plant. "Restricted Use" pesticides should only be used with the required supervision and "General Use" pesticides should only be used by personnel with the required training. Every effort should be made to prevent the likelihood of contamination of food and food contact surfaces with pesticides. All pesticides and all pesticide application equipment must be clearly identified with labels and stored in a protected, locked area far removed from food processing areas and storage areas for raw materials, ingredients, packaging materials, cleaning materials and products.

4.11.4   Monitoring for effectiveness

The pest control program should be monitored for effectiveness on a continuous basis. Whenever there is evidence of pests or pest activity at the pest control devices or at any location in a food plant, the source of the pest should be identified and eliminated as soon as possible. In addition, the pest control program should be reviewed to determine that the preventive aspects of the program are effective.

4.11.5   Pest control records

A food plant should maintain records as part of its pest control program. These should include the reports of the scheduled monitoring of pest control devices along with evidence of pests and pest activity and records relating the use of any pesticide for pest control within the plant.

4.12 HACCP prerequisite program equipment

This prerequisite program covers activities directed at design, construction, installation, performance, maintenance, and use of equipment in a food plant. It also includes the calibration of equipment used for monitoring and measuring parameters at any point in a process of detection, elimination, control, or prevention of food safety hazards, and for measuring product characteristics that are indicators of the safety of a product. The cleaning and sanitation of equipment is covered under the prerequisite program *Sanitation and cleaning*. Equipment used for storage of materials and products is covered under the prerequisite program *Transportation, receiving, storage, and shipping*.

The particular types of equipment that are used in a food plant depend on the specific type of products that are processed. The performance of equipment should ensure that the safety or quality specifications of a food can be achieved. In this prerequisite program, the primary considerations are to ensure that the equipment are capable of processing products that meet the safety and quality requirements, while at the same time the equipment must not be a source of contamination of the product.

This prerequisite program covers the items addressed in *Subpart C—Equipment* of the *Current Good Manufacturing Practices (Part 110, Title 21, CFR)* and in *4.3 Equipment* in *Establishment: Design and Facilities* of the *General Principles of Food Hygiene (Codex Alimentarius Commission)*. Table 4.6 lists the topics which are covered by this prerequisite program.

Table 4.6 Topics Covered by Prerequisite Program
Equipment

Processing equipment
– Design and installation
– Food-contact surfaces
– Installation
– Maintenance program
– Maintenance personnel
Handling equipment
Storage equipment
Monitoring and monitoring equipment
– Calibration

Relationship to ISO 9001: The requirements of sections *6.3 Infrastructure, 7.5.1 Control of production and service provision,* and *7.6 Control of monitoring and measuring devices* address the topics covered by this prerequisite program.

4.12.1   *Processing equipment*

- *Design and construction:* All equipment should be suitably designed and constructed to ensure that the specific requirements of the process can be achieved and the required maintenance, inspection, and cleaning can be readily undertaken. In addition, during the operation and use of equipment, there should be no contamination of product from the equipment itself, and there should be no unacceptable accumulation of any material (e.g., dust, metal fragments, oil, water, product) that is likely to be a source of contamination. In some food processing operations, a government regulatory agency specifies the design and construction requirements of the equipment that are used.

- *Food-contact surfaces*: The food-contact surfaces of equipment should be made of nontoxic material and should not be corroded or damaged in any way during normal operations, or when in contact with raw materials, products, and cleaning materials. The seams on food contact surfaces should be smooth so as to prevent accumulation of product and to facilitate cleaning and sanitizing.

- *Installation*: All processing equipment should be installed in a manner that will facilitate its operation, and the cleaning and maintenance of both the equipment and its immediate surroundings. After any equipment has been installed, it should be inspected and approved for use before it is put into regular operation. This inspection and approval should confirm that the equipment is capable of performing the intended operation and that the equipment is not a source of contamination.

- *Maintenance program*: In addition to the equipment cleaning and sanitizing that are covered in the prerequisite program *Sanitation and Cleaning*, there should be an equipment maintenance program to ensure that the equipment is always operating as is intended, is meeting the requirements of the process, and is not a source of contamination. This program should cover (a) the routine maintenance such as cleaning, inspection, servicing and lubrication, (b) repairs and unscheduled maintenance resulting from equipment break-down during regular operations and scheduled and planned preventive maintenance based on the equipment manufacturer's guidelines or on the conditions and period of operations. If any equipment has been disassembled during repair or maintenance, it should be inspected and approved before it is returned to regular use. Lubricants used for maintenance of food processing equipment should be food grade and approved for use by the appropriate government regulatory agency. Excess lubricant from servicing of equipment or lubricant that accumulates during operation of any equipment should be removed and prevented from coming into contact with food.

- *Maintenance personnel*: Personnel who carry out equipment maintenance activities should be aware of the practices to be followed to ensure that contamination of product does not occur as a result of equipment maintenance.

4.12.2 Handling equipment

Equipment such as forklifts and hand-jacks that are used to handle and move pallets with bags, cartons, or containers with materials and products at a food plant, should be maintained so as to prevent damage to or contamination of foods.

4.12.3 Storage equipment

All storage equipment should be maintained to ensure that the materials and products are stored under the appropriate conditions to prevent deterioration or contamination of foods. This is particularly applicable to equipment that store foods at controlled temperatures (e.g., freezers, refrigerators, storage tanks).

4.12.4 Monitoring and measuring equipment

- *Calibration*: There should be a calibration program for equipment that is used for monitoring and measurement of safety- and quality-related characteristics of raw materials, ingredients, and products, and of processing parameters. The calibration program should include the identification of all equipment that is in the program, the schedule for calibrating the equipment, the person responsible for calibration, and the procedures to be followed for performing the calibration. Calibration records should be kept for the equipment that is included in the program. The calibration program should also include control of all reference standards including chemical reagents that are used for calibrating equipment.

- *Monitoring equipment*: All monitoring equipment including devices for detecting the presence of metal fragments and sensors for examination of the integrity of can seams and package seals should be calibrated

- *Measuring equipment*: All measuring equipment should be calibrated. This includes devices for measuring temperature of thermal processes such as cooking, sterilization, and pasteurization, temperature of cold storage rooms such as coolers and freezers, equipment for measuring quantities of ingredients which act as preservatives and equipment for measuring safety, and quality-related product characteristics such as pH, water activity, viscosity, color, and net weight.

4.13 HACCP prerequisite program transportation, receiving, storage, and shipping

This prerequisite program covers the activities directed at transportation and receiving of all materials to be used in the processing and packaging operations, storage and warehousing of all materials and products at the food establishment, and shipping and transportation of all foods to the point of delivery.

In this prerequisite program, the primary consideration is the prevention of contamination, cross-contamination and deterioration of products by (a) control of incoming raw materials, ingredients, packaging materials, and processing aids from the time they arrive at a food plant, their subsequent inspection, acceptance and storage until they are utilized, and (b) control of the storage of semi-finished and finished products from the time they are produced, their subsequent move to designated storage areas or containers, their shipment, delivery, and distribution to their point of use or sale.

This prerequisite program addresses some topics in the section on *Processes and controls*, and *Warehousing and distribution* in *Subpart E—Production and Process Controls* in the *Current Good Manufacturing Practices* (Table 4.1) and in *5.3 Incoming Material Requirements* in *Control of Operations* and *Transportation* of the *General Principles of Food Hygiene* (Table 4.1). Table 4.7 lists the topics that are covered by this prerequisite program.

Table 4.7 Topics Covered by the Prerequisite Program
Transportation, Receiving, Storage, and Shipping

Transportation/receiving
– Receiving location
– Transport vehicles
– Verification of received materials
Handling and storage
– Raw materials, ingredients and packaging materials
– Nonfood chemicals
– Half-finished and finished products
– Storage conditions
– Stock rotation/First-in-First-out (FIFO)
Transportation shipping
– Finished products
– Transport vehicles
– Delivery

Relationship to ISO 9001: The requirements of sections *7.4.3 Verification of purchased product* and *7.5.5 Preservation of product* address the topics covered by this prerequisite program.

4.13.1 Transportation and receiving

- *Receiving location*: The location for receiving raw materials, ingredients, packaging materials, and nonfood materials at a food plant should be separate from food processing areas to prevent any likelihood of cross-contamination of products. The activities at the receiving location should be limited to inspection and, if necessary, sampling of the incoming materials. Materials received should be transferred to their designated storage area and not stored at the receiving location.

- *Transport vehicles*: All vehicles including containers, railcars, and tankers, that deliver raw materials, ingredients, and packaging materials to a food plant should be inspected to ensure that the sanitary conditions of the vehicle are satisfactory and there was no potential for tampering, contamination or infestation of materials during the delivery. This inspection should be carried out before or during the unloading of the vehicle, and should address the actual state of the materials in the vehicle, including the integrity of packages and containers, ceiling, walls and floor, the presence of any abnormal odor, the presence of hazardous materials or pests, the verification of seals and seal numbers, any requirement for temperature control, and any other specific requirements associated with a particular delivery. Any unacceptable condition relating to the sanitary conditions of the transport vehicle, or damage to packages or containers, or evidence of tampering should be addressed before the materials are transferred to their designated storage areas in the plant. A record of the inspection of the transportation vehicle and the incoming materials at the receiving location should be kept.

- *Verification of received materials*: Raw materials and ingredients received at a food plant should be verified to determine that they are suitable for their intended use. Raw materials and ingredients that contain biological, chemical, or physical hazards that would not be reduced to an acceptable level by the processing operations should not be used. Specifications or other requirements relating to safety and quality should be established for raw materials, ingredients, and packaging materials and should be used as the basis for acceptance when this verification is performed. Records of verification of received materials should be kept. Raw materials and ingredients that are covered by HACCP plans should not be used until all the HACCP plan requirements are met.

4.13.2   Handling and storage

- *Raw materials, ingredients, and packaging materials*: The handling of received materials during unloading and discharge of the vehicles, and the transfer to designated storage areas, should ensure that there is no damage or contamination. Raw materials should be stored separately from finished products to prevent cross-contamination. All received materials should be appropriately identified and transferred to the designated storage area immediately after unloading.

- *Nonfood chemicals*: All nonfood chemicals such as cleaning and sanitizing agents, and pesticides, should be stored in secured, segregated areas to prevent contamination of other materials, products, or equipment.

- *Semifinished and finished products*: Semifinished and finished products should be appropriately identified or tagged and transferred to their designated storage areas immediately after processing. All necessary precautions should be taken to prevent contamination or cross-contamination during handling and storage of these products.

- *Storage conditions*: The conditions for storage of all received materials and all products, including product to be re-worked, should prevent and protect from damage, deterioration, contamination, cross-contamination or pest infestation during storage. Temperature (e.g., refrigeration, freezing) and humidity conditions for storage should be established, if necessary. All storage conditions, including pest control requirements should be monitored, and records of this monitoring should be kept. Storage conditions that are covered by HACCP plans must meet all the HACCP plan requirements.

- *Stock rotation/first-in-first-out (FIFO):* The use of stored materials and ingredients, and the shipment of finished products should be controlled in a manner that older or first-received (first-in) materials and ingredients are used first (first-out), and older or first-produced products (first-in) are shipped first (first-out). This rotation of stock prevents extended storage of materials and products that can result in their deterioration and pest infestation.

4.13.3 Transportation/shipping

- *Finished products*: Only products that have met all the food safety requirements, including all the HACCP plan requirements, the product specifications, and packaging and labeling requirements relating to product safety, and which have been approved for shipment to customers, should be prepared for transportation and delivery.

- *Transport vehicle*: All product transportation vehicles, including tankers, containers and railcars should be inspected before product is loaded for delivery of products. This inspection should cover the cleanliness and sanitary conditions of the vehicle and its accessories, its ceiling, walls, and floor, capability to meet any requirements for temperature and for seals, and any restriction relating to the previous use of the vehicle or previous load carried by the vehicle. A record of this inspection should be kept.

- *Delivery*: Adequate measures should be taken during transportation to prevent damage, deterioration, contamination of, or tampering with products that are to be delivered to customers. Any temperature requirements during transportation must be met and the temperature should be recorded. Locks and seals on delivery vehicles must be maintained during transportation. If seals have to be removed for inspection by customs agencies at border crossings, the seals must be replaced and this should be recorded.

4.14 HACCP prerequisite program traceability and recall

This prerequisite program addresses the requirements for identifying and tracing all raw materials, ingredients, and products, including all semi-finished, finished, reworked, or recycled products, and the procedures to be followed for conducting a recall of a food that has reached the marketplace, should this become necessary. In the food industry, it is quite common for foods to be recalled from the marketplace. This occurs in spite of the enforcement of and compliance with regulatory requirements of GMPs in food establishments and the use of government mandated HACCP system requirements in some food industry sectors.

A recall of a food can take place whenever it has been determined that there is an unacceptable health hazard associated with the food that has reached the consumer marketplace, or some aspect of the food violates the laws or regulations that govern the product. A recall can be initiated without any evidence that the health of a consumer of the food has been affected. A food can contain an unacceptable health hazard as a result of deficiencies during the manufacture of the product. For some types of foods, contamination can take place during storage, distribution, handling, and retailing.

The accurate recorded identification of the usage and movement of raw materials, ingredients, and products, including recycled or reworked products, at all stages of processing, handling, storage, and distribution of a food, are essential for the traceability of a food.

This prerequisite program covers some of the sections included in *Subpart C—Recalls (Including Product Corrections)—Guidance on Policy, Procedures,* and *Industry Responsibilities* of *Part 7— Enforcement Policy (Title 21, CFR)* and in *Recall Procedures* in *Control of Operations* and *Product Information and Consumer Awareness* of the *General Principles of Food Hygiene* (Table 4.1).

Relationship to ISO 9001: The requirements of section *7.5.3 Identification and traceability* address the topics covered by this prerequisite program.

4.14.1   Identification

All raw materials, ingredients, and products, including intermediate, semi-finished, pre-finished, finished, recycled, reworked, pre-packaged, and packaged products should be identified and the identification recorded. The identification of a food to be shipped from a food plant should cover the identity of the food (e.g., product name, recipe, product code) and the time or period of production (e.g., production date, day code, lot number, batch number). In addition, all packaged foods must conform to the applicable ingredient identification requirements of government labeling regulations. All raw materials, ingredients, processing aids, packaging materials, and the processing and storage equipment used for the production of a particular food, should be identified and the identification recorded to ensure that if it becomes necessary, the history of the food can be traced.

The locations of storage and addresses of places to which a food has been distributed for use by customers and for sale to consumers should be identified in the appropriate shipping and distribution records, which should be kept for a period that exceeds the life of the food in the marketplace.

4.14.2   Traceability

There should be a program for traceability for every food that is manufactured and shipped from a food plant. This program should cover forward traceability (for use of raw materials and ingredients) and backward traceability (for finished products). The traceability program should be based on the use of the relevant identification records and should cover all foods that are in inventory in all warehouses, foods that have been shipped from a food plant and its warehouses, foods delivered to customers, and foods that have been sold to consumers.

4.14.3   Recall procedure

A written recall procedure should be developed to ensure that if a food which is known to present an actual or potential unacceptable health risk to the consumer has entered the food distribution and retail chain, the food can be retrieved quickly and completely. The recall procedure must meet the requirements or guidelines for recall issued by government regulatory agencies, it should describe a precise and well-coordinated set of activities that can be executed very quickly and efficiently. The recall procedure should be tested periodically to determine its effectiveness and should be based on identification records and the traceability program. The procedure should include, as a minimum, the following information:

- The people, including their alternates if they cannot be contacted, who will be responsible for conducting the recall, and the people who should be available to provide information, along with the specific responsibilities of each individual
- The details for contacting the appropriate senior management or other personnel with assigned responsibilities, including contact information for these individuals during periods outside of regular working hours
- The details for contacting and communicating with the appropriate government regulatory agency, with all customers who would have received the food being recalled, and with the news media
- The person designated to be the liaison with officials from a government regulatory agency
- The records and other information that should be retrieved during a food recall. The required information includes (a) all identification information related to the food, including its common and brand names, item number, lot number, batch number, best before or expiration date, product code, UPC code, packaging materials and packaging format, packaging container size, (b) a list of all locations, with addresses, telephone numbers and contact information, to which the food has been shipped, including all warehouses for its storage, all customers, including distributors, retailers, food service institutions and restaurants, and (c) identification of all raw materials, ingredients, processing aids, packaging materials, and processing and storage equipment used for the manufacture of the food, and the processing conditions and storage conditions used for the food
- The procedure to be followed to determine which production lots or batches of the food are affected to ensure that the scope of the recall targets all the affected food, but only the affected food.

A food company should have a permanent crisis-management team that has responsibility for food recall, as well as other critical food safety concerns (e.g., food tampering, and other critical safety concerns, environmental safety, employee safety).

4.15 HACCP prerequisite program chemical control

This prerequisite program addresses the control of the various chemical hazards that are used in food plants and in the processing of foods that are not covered by CCPs of HACCP plans (Chapter 5). These chemical hazards include some permitted food additives, foods that are known allergens, cleaning and sanitizing chemicals, pest control chemicals, and chemicals used for equipment maintenance. The control of many of these chemicals can be achieved by the other prerequisite programs described earlier. For allergens, an allergen control program should be developed to control allergens that are not controlled as chemical hazards by CCPs of HACCP plans (Chapter 5). This program should be devoted to preventing allergen contamination or cross-contamination of foods which do not contain allergenic ingredients.

4.16 Other HACCP prerequisite programs

In addition to the activities that are covered in sections 4.8 to 4.15, there are several activities in food processing plants that can be considered as prerequisite programs for HACCP. These include activities that relate to food safety customer complaints, written specifications for materials and products, and exercising control over suppliers of materials and services. In general, food companies cover these activities in their quality programs or quality systems (Chapter 3)

4.17 Production and process control

The activities at food processing plants should include process controls. At plants that operate with HACCP systems, the process controls should address the processing activities that are not covered by the CCPs of HACCP plans. At plants which do not have HACCP systems, the process control activities should address all the food processing requirements for food safety. In general, food companies include process control activities in their quality programs or quality systems (Chapter 3).

4.18 Monitoring of GMPs and prerequisite programs

The GMPs and prerequisite programs should be monitored continuously to determine that they are effective for fulfilling the applicable food safety requirements. Auditing and inspection activities should be performed to demonstrate effectiveness. In general, food companies include audits and inspections as part of their quality programs and quality systems (Chapter 3).

References

Current Good Manufacturing Practice in Manufacturing, Packing, or Holding Human Food, Codes of Federal Regulations, Title 21, Part 110, Office of the Federal Register, National Archives and Records Administration. Washington, D.C., www.access.gpo.gov/cgi-bin/cfrassemble.cgi?title=200221.

Enforcement Policy Recalls (Including Product Corrections)—Guidance on Policy, Procedures, and Industry Responsibilities. Code of Federal Regulations, Title 21, Part 7, Subpart C, Office of the Federal Register, National Archives and Records Administration, Washington, D.C., www.access.gpo.gov/cgi-bin/cfrassemble.cgi?title=200221.

Food Safety Enhancement Program, FSEP Implementation Manual, Canadian Food Inspection Agency Food Safety Enhancement Program, www.inspection.gc.ca/english/ppc/psps/haccp/manu/manue.shtml.

Hazard Analysis and Critical Control Point Systems, Code of Federal Regulation, Title 21, Part 120, Office of the Federal Register, National Archives and Records Administration, Washington, D.C., www.access.gpo.gov/cgi-bin/cfrassemble.cgi?title=200221.

Hazard Analysis and Critical Control Point Systems, Meat and Poultry HACCP Regulation, Code of Federal Regulation, Title 9, Part 416, Office of the Federal Register, National Archives and Records Administration. Washington, D.C., USA. www.access.gpo.gov/cgi-bin/cfrassemble.cgi?title=200221.

Hazard Analysis and Critical Control Point Principles and Application Guidelines, Adopted August 14, 1997 by National Advisory Committee on Microbiological Criteria For Foods, *J. Food Prot.*, 61(9), 1246–1259, 1998.

International Standard ISO 9001:2000 quality management systems—requirements. 3rd ed. ISO 2000, Geneva.

Procedures for the Safe and Sanitary Processing and Importing of Fish and Fishery Products. Seafood HACCP Regulation, Code of Federal Regulations, Title 21, Part 123, Office of the Federal Register, National Archives and Archives Administration, Washington, D.C., www.access.gpo.gov/cgi-bin/cfrassemble.cgi?title=200221.

Recommended International Code of Practice, General Principles of Food Hygiene, Codex Alimentarius Commission, Joint FAO/WHO Food Standards Programme, 1997, Rome, www.codexalimentarius.net/.

The Quality Auditor's HACCP Handbook. ASQ Food, Drug, and Cosmetic Division, 2002, ASQ Quality Press, Milwaukee.

chapter five

The HACCP system for food safety

5.1 Introduction

This chapter deals with the hazard analysis critical control point (HACCP) system, its principles and its practices for addressing the safety of foods produced in food processing establishments. At a food plant that operates with the HACCP system, the system is comprised of the HACCP prerequisite programs, which apply to the entire establishment, and the HACCP plans, which apply to the foods produced at the establishment. HACCP prerequisite programs are covered in Chapter 4. The HACCP plan for a food is developed on the basis of the seven principles of HACCP; the development of an HACCP plan is covered in this chapter.

The HACCP system, with its prerequisite programs is essentially a food safety management system. Many of the practices of the HACCP system are similar to the activities of a quality management system, which has been covered in Chapter 3; however, the practices of the HACCP system are devoted specifically to the safety requirements of foods. Since food safety is a specific component of food quality, a food company's HACCP system can be integrated within its quality system activities.

5.2 Evolution of the HACCP system

The evolution of the HACCP system during the second half of the twentieth century, from its roots in the U.S. space program to its present use for consumer foods worldwide, can be traced through several milestones. In addition, the evolution of the HACCP system can be traced through a series of formal recognition and adoption activities by the food industry, government regulatory agencies, and national and international scientific and professional organizations.

5.2.1   Origin of HACCP

Most experts agree that the development and initial use of an HACCP approach for food safety can be traced to a joint effort of the U.S. National Aeronautic and Space Administration (NASA), the U.S. military, and the Pillsbury Company in the late 1950s and early 1960s. One objective of this collaboration was to develop a strategy that would ensure that foods required for the space program were free of any unacceptable health risk. The Pillsbury Company, in conjunction with NASA and the U.S. Army Natick Laboratories, pioneered the development of the HACCP approach for preventing unacceptable levels of food safety hazards. The need for this preventive approach resulted from the recognition that the approach in use at that time for food safety was based primarily on inspection and testing of foods, and was neither practical nor effective for ensuring that foods for the space program were free of unacceptable health hazards. This marked the beginning of the use of the HACCP system for addressing food safety.

5.2.2   The HACCP system for consumer foods

After the initial success with foods for the space program, the Pillsbury Company pioneered the development of the HACCP system for food safety in the manufacture of consumer foods in its food processing plants. The Pillsbury Company announced the use of the HACCP system for consumer foods in the early 1970s, and subsequently played a leading role in providing expertise, information and training to the food industry and to government regulatory agencies in the U.S. This resulted in the general acceptance of the HACCP system by food manufacturers and government regulatory agencies; however, the actual adoption and use of the HACCP system by food manufacturers was not extensive initially.

5.2.3   Recognition of the HACCP system

The recognition of the HACCP approach in 1985 by the National Academy of Sciences as a preventive approach for ensuring the microbiological safety of foods, generated considerable renewed interest in the use of the HACCP system. This recognition of the HACCP approach was followed by substantial contributions from the U.S. National Advisory Committee on Microbiological Criteria for Foods (NACMCF) toward the development of the HACCP system. The work of NACMCF resulted in several substantial publications on the HACCP system; the 1997 edition of NACMCF's publication *Hazard Analysis and Critical Control Point Principles and Application Guidelines* is used extensively as a primary reference document on HACCP.

5.2.4   International recognition of the HACCP system

In 1987, the International Commission on Microbiological Specifications for Foods (ICMSF) of the World Health Organization (WHO) endorsed the use of the HACCP approach. More extensive international recognition of HACCP emerged in 1991 when the Codex Committee on Food Hygiene prepared a draft report on HACCP for Codex Alimentarius member countries. Essentially, the Codex Committee recommended that the HACCP system be accepted as the basis for an internationally recognized approach for addressing food safety. The final version of the Codex Alimentarius HACCP system was later incorporated into the Codex Alimentarius *Basic Texts on Food Hygiene.*

5.2.5   Use of HACCP in government regulations

The U.S. Food and Drug Administration first formally used the HACCP approach in 1973 for the government regulation of low acid canned food. By the late 1980s, HACCP had received broad-based endorsement and acceptance by the food industry, by government regulatory agencies in many parts of the world, and by the scientific community in general. With this favorable climate, several countries embarked on the road toward development of national regulatory food safety programs based on HACCP. In the U.S. and Canada in the early 1990s, the seafood sector was the first to come under voluntary HACCP-based inspection programs. Subsequently, mandatory HACCP-based inspection programs have been developed in the U.S. for the seafood sector in 1995, the meat and poultry sector in 1996, and the processed juice sector in 2001. There is every indication that mandatory HACCP-based inspection programs will become more widespread, not only for other food industry sectors in the U.S., but also in many other countries.

5.3   The seven HACCP principles

The HACCP system is based on a universally recognized set of seven principles that are used for the development of an HACCP plan for a food. These principles reflect a framework that was developed on the basis of a combination of recognized, science-based, food safety considerations and quality systems characteristics. This integration of basic food safety principles with the quality systems approach has been an important factor in the widespread recognition of the HACCP principles by food quality professionals.

The universally recognized Seven Principles of HACCP are:

- Principle 1: Conduct hazard analysis
- Principle 2: Determine critical control points
- Principle 3: Establish critical limits
- Principle 4: Establish monitoring procedures
- Principle 5: Establish corrective action procedures
- Principle 6: Establish verification procedures
- Principle 7: Establish record-keeping and documentation procedures

5.4   The HACCP system and HACCP plans

Within a particular food processing establishment, the scope of the HACCP system will depend on the number of different food products or different categories of similar food products that are produced. In the case where a single product is produced at a food establishment, the HACCP system is comprised of a single HACCP plan for the product that is produced by a specified process, along with the HACCP prerequisite programs that cover the entire establishment. In food establishments that produce several products, it should be determined whether similar products can be grouped into a family of products that could be covered by a single HACCP plan. In an establishment at which different products are produced using different raw materials and ingredients, different processes and process parameters, and different processing equipment, it is very likely that the HACCP program for the establishment will consist of several HACCP plans, with an HACCP plan for each product or each family of products.

5.5   Development and implementation of an HACCP plan

The widespread recognition of HACCP is reflected not only in its science-based, preventive approach, but also in the requirements and guidelines for the development and implementation of an HACCP plan. A model consisting of a sequence of twelve steps for developing and implementing an HACCP plan for a particular food product has been adopted by the Codex Alimentarius and is referred to as the *Logic Sequence for the Application of HACCP*. This set of twelve steps includes five preparatory steps that provide guidance and which are also considered necessary preliminary steps by the USNACMCF, followed by seven steps that address the seven principles of HACCP. The worldwide acceptance of the Codex Alimentarius model has resulted in a consistent format for developing and implementing HACCP plans in food establishments around the world, and has contributed to the recognition of HACCP systems to address food safety issues in bilateral and international trade.

5.6   The Codex Alimentarius Logic Sequence for the Application of HACCP

The twelve steps in the Codex Alimentarius Logic Sequence for the Application of HACCP are as follows:

- **Step 1:** Assemble an HACCP team.
- **Step 2:** Describe the food product that the HACCP plan will address.
- **Step 3:** Identify the intended use of the food product.

- **Step 4:** Construct a flow diagram of the process that is used to produce the food product.

- **Step 5:** Conduct an on-site verification of the process flow diagram.

- **Step 6:** Conduct a hazard analysis of (a) all raw materials and ingredients and (b) each step (in the process flow diagram) used for preparation of the food product (*HACCP Principle 1*).

- **Step 7:** Determine which (a) raw materials and ingredients, and (b) process steps, will be critical control points at which unacceptable hazards identified in *Step 7*, will be controlled (*HACCP Principle 2*).

- **Step 8:** Establish critical limits or tolerances for each of the critical control points identified in *Step 7* (*HACCP Principle 3*).

- **Step 9:** Establish monitoring procedures for each of the critical control points identified in *Step 7* (*HACCP Principle 4*).

- **Step 10:** Establish corrective action procedures to be followed when monitoring of the critical control points reveals that the established critical limits have been exceeded or have not been met (*HACCP Principle 5*).

- **Step 11:** Establish verification procedures to confirm and provide confidence that (a) the critical control points are being monitored effectively and are under control, and (b) the HACCP plan for the product is operating effectively (*HACCP Principle 6*).

- **Step 12:** Establish record-keeping and documentation procedures for records and documents that are required by the HACCP plan (*HACCP Principle 7*).

Step 1 establishes the HACCP team that will perform *Steps 2 to 12*. Each of the twelve steps of the Codex Alimentarius Logic Sequence is described in detail in the following sections; for each step, the relationship to the requirements of the *ISO 9001:2000 Quality management system* is indicated.

5.6.1   *Step 1 — assemble an HACCP team*

An HACCP team is a group of a food company's employees who should be assembled and given the responsibility by management to develop and implement an HACCP system for the company's establishment at which the food product is produced. Essentially, the team carries out *Step 2* to *Step 12* of the Codex Alimentarius Logic Sequence for the Application of HACCP. The team, which commonly consists of four to eight people, should be composed of personnel who can contribute knowledge in quality assurance, quality control, food microbiology, food processing, GMPs, and equipment maintenance. Personnel who are responsible for ongoing activities in inspection, testing, production, cleaning, and sanitation can also be included in the HACCP team. A member of the HACCP team should serve as the HACCP coordinator who acts as the leader of the team and ensures that all the requirements for a recognized HACCP system are covered. It is essential for the HACCP coordinator to have comprehensive training in the development, implementation and maintenance of an HACCP system. The other members of the team should also receive some level of training in the HACCP system. If the food company's products are covered by an HACCP-based inspection program of a government regulatory agency, then the HACCP team must be aware of the specific requirements of the regulatory agency's program.

The company's senior management must demonstrate its commitment to the development, implementation, and maintenance of the HACCP system. The HACCP team must receive support from its management, which must provide the resources required. Management must ensure that all members of the team obtain the required training and are available to work as part of the HACCP team. In addition, the appropriate training should be provided to all employees whose work will involve the use of the HACCP plans that are developed and implemented.

Relationship to ISO 9001: The following clauses of *ISO 9001:2000* can be related to *Step 1* of HACCP system development, implementation, and maintenance, *5 Management responsibility (5.1 Management commitment, 5.2.2 Responsibility and authority), 6 Resource requirements (6.1 Provision of resources, 6.2 Human resources).*

5.6.2   Step 2 — describe the food product which the HACCP plan will address

The important characteristics relating to the safety of the product must be clearly described by the HACCP team. This information will be used by the team in the identification and analysis of all hazards *(Step 6, HACCP Principle 1)* associated with all aspects of preparation of the product.

The product characteristics that should be described include the following:

- The product name, including all alternate or common names for the product (or the name of a family of similar products)
- The composition of the product, or the physical or chemical properties (e.g., pH, Aw, preservatives, presence of allergenic ingredients) that must be controlled to ensure the safety of the product
- The packaging of the product, including the package unit (e.g., can, bag, case), the packaging material (e.g., foil, plastic, paper), and packaging conditions (e.g., modified atmosphere packaging)
- The shelf-life of the product and any required storage temperature (e.g., refrigerated, frozen) and humidity conditions
- The labeling instructions to customer or consumer for handling, storage (e.g., refrigerated, frozen) and use (e.g., cooking time and temperature) of the product

- Any special conditions for distribution of the product (e.g., refrigeration or freezing during shipping)
- The actual use of the product (e.g., ready-to-eat, heat before consumption, industrial use with further processing)
- The sale of the product (e.g., retail to consumers, to institutions, to industrial customers).

Relationship to ISO 9000: The following clauses of *ISO 9001:2000* can be related to *Step 2* of the HACCP plan development and implementation: *7 Product realization (7.2.1 Determination of requirements related to the product, 7.2.2 Review of requirements related to the product).*

5.6.3   Step 3 — identify the intended use of the product

The normal or common use of the product must be known in order for the hazard analysis to be done *(Step 6; HACCP Principle 1).* Therefore, this step should establish where and by whom the product will be used (e.g., food service or institutional use, industrial use, or household use by the general consumer). Certain segments of the population (e.g., elderly persons, pregnant women, infants, individuals whose immune system is compromised) are at higher risk to certain biological hazards and chemical hazards; therefore the use of the product by these groups need to be determined. Some of the information required in *Step 3* can be obtained at *Step 2* and it is possible for these two steps to be combined.

Relationship to ISO 9000: The following clauses of *ISO 9001:2000* can be related to *Step 3* of the HACCP plan development and implementation: *7 Product realization (7.2.1 Identification of customer requirements).*

5.6.4   Step 4 — construct a process flow diagram for the product

The HACCP team, with assistance from personnel who are familiar with the process, should construct a process flow diagram that shows a simple but logical step-by-step outline of the process from which the product will be obtained. The process flow diagram should identify all key steps from receiving of raw materials and ingredients for preparing of the product; all handling, sorting, preparation and storage of raw materials and ingredients; all processing treatments including steps at which there are filters, screens, magnets, and metal detectors; all packaging, labeling, and storage steps through to shipping of the product from the establishment. Any rework material to be used in the product should also be identified in the flow diagram. The process flow diagram is extremely useful for conducting the hazard analysis and for the critical control point determination *(Steps 6 and 7, HACCP Principles 1 and 2).* Figure 5.1 shows a simple, generic process flow diagram for packaged food product.

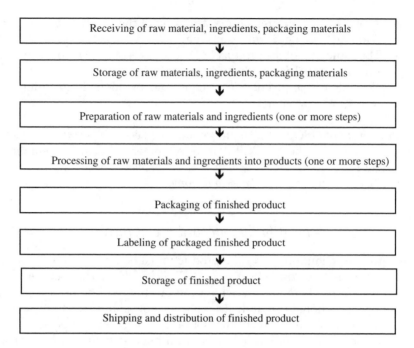

Figure 5.1 A generic process flow diagram for a packaged food product.

Relationship to ISO 9000: The following clauses of *ISO 9001:2000* can be related to *Step 4* of the HACCP plan development and implementation: *7 Product realization (7.1 Planning of realization process, 7.5 Production and service provision).*

5.6.5   *Step 5 — verification of the process flow diagram*

The HACCP team, with assistance from personnel who actually operate the process, should verify the process flow diagram prepared in *Step 4*, to establish that the diagram accurately represents the actual activities and operations used to prepare the product. This is done by observing each step of the process, from receiving of raw materials to shipping of finished product, as each activity and operation identified on the process flow diagram actually takes place. Based on the results of this observation, the process flow diagram should be modified as required. Alternatively, if a necessary step for preparation of the product is not in place, this step should be introduced into the process. The accuracy of the process flow diagram in representing the actual process required for the preparation of the product is essential for the successful development of the HACCP plan.

Relationship to ISO 9000: The following clauses of *ISO 9001:2000* can be related to *Step 5* of the HACCP plan development and implementation: *7 Product realization (7.1 Planning of product realization, 7.5. Production and service provision).*

5.6.6   Step 6 — conduct a hazard analysis (HACCP Principle 1)

After *Steps 1 to 5* have been completed, the HACCP team can then proceed to develop the HACCP plan for the product. *HACCP Principle 1* hazard analysis must be performed at this stage. The hazard analysis can be an extensive exercise for the HACCP team, since it addresses the three categories of hazards — biological, chemical, and physical. The hazard analysis should be carried out in two stages; the first stage is *hazard identification,* which is followed by *hazard evaluation.* In addition, the control measures for the hazards that need to be addressed in the HACCP plan should be identified at the completion of the hazard analysis. Table 5.1 gives examples of biological, chemical and physical hazards that have been identified in some common processed foods. This is not a complete list of all the potential hazards in processed foods.

5.6.6.1   Hazard identification

The first stage of hazard analysis is the identification of known potential hazards that are likely to be associated with the product; this covers the ingredients, raw materials, and contact packaging materials used in preparation of the product; every aspect of each step of the process; and the environment used for the preparation of the product. If all the potential hazards are not identified, the HACCP plan that is developed might be

Table 5.1 Examples of Some Biological, Chemical, and Physical Hazards in Foods

Biological hazards: Pathogenic microorganisms, e.g.,	
Clostridium botulinum	*Salmonella spp.*
Pathogenic *Escherichia coli*	*Listeria monocytogenes*
Staphylococcus aureus	*Clostridium perfringens*
– Viruses	– Parasites
Chemical hazards:	
– Antibiotics	– Cleaning and sanitizing chemicals
– Hormones	– Regulated food additives
– Drugs	– Allergens
– Agricultural residues	– Toxicants
– Industrial pollutants	
Physical hazards:	
– Glass	– Stones
– Wood	– Bones
– Hard plastic	– Filth
– Metal	– Personal articles

inadequate to address the safety of the product. The NACMCF has compiled a list of examples of issues that should be considered during a hazard identification exercise:

- The nature of the product and the product characteristics that have been established during the completion of *Step 2*
- The safety record of the product and the known hazards associated with the product
- The normal microbial content of the product and changes in the microbiological population during storage and handling of the product
- The hazards that are likely to be present in the product if product characteristics such as pH, composition, water activity, and preservatives are not controlled
- The various raw materials, ingredients, and packaging materials used to prepare the product
- The activities, operations, and personnel involved at each of the steps listed in the process flow diagram *(Step 4)*, including the equipment that is used and the prevailing environmental conditions at the time the product is produced and stored
- The intended use of the product by customers, consumers, or particular segments of the population *(Step 3)*

During the hazard identification exercise, the HACCP team should consult appropriate sources of information to determine the known product hazards for which the HACCP plan is developed, as well as known hazards associated with the various raw materials, ingredients, and the processing operations used for the product. A useful database *(Reference Database for Hazard Identification)* has been developed by *Agriculture and Agri-Foods Canada (1995)*. In addition to reference information, valuable information on potential hazards can be obtained by closely observing the process as it is operating. In order for the hazard identification to be successful, it is essential that the HACCP team acquire detailed knowledge of potential hazards associated with every aspect of each step of the process that produces the product for which the HACCP plan is developed.

5.6.6.2   Hazard evaluation

The second stage of the hazard analysis is the evaluation to determine which of the hazards that have been identified in the first stage of the hazard analysis are significant, and therefore need be addressed in the HACCP plan for the product. The identified hazards must be considered for their *likely occurrence*, and the *severity* of the risk that they present if they are not controlled. In assessing the *likely occurrence* of a hazard, the HACCP team needs to examine published information and data, and previous experiences on occurrence of the hazard in the product. In assessing *the severity* of the risk presented by a hazard, the HACCP team must address the severity of health consequences to consumers of the product, if the identified hazard is not controlled. In carrying out this stage of the hazard analysis, the HACCP team should take the following into account:

- The methods and procedures of preparation, handling, storage, and distribution of the product
- The effects of both short-term and long-term exposure of the product containing the hazard to the health of consumers
- The people, including any particular groups (e.g., infants, pregnant women, the elderly) who are likely to consume the product and the effect the hazard could have on the health of these individuals
- Information and data from previous reported incidents involving the occurrence of the hazard in the product, including the impact of the hazard on the health of consumers.

5.6.6.3 Identification of control measures

As part of the hazard analysis the HACCP team must determine whether control measures for the hazards that need to be addressed in the HACCP plan are present in the process. If control measures for the hazard do not exist in the process, modification of the process may be required in order to institute a control measure, or some alternate measure should be used to address the hazard. Table 5.2 provides some examples of control measures for biological, chemical and physical hazards.

Relationship to ISO 9000: The following clauses of *ISO 9001:2000* can be related to *Step 6 (HACCP Principle 1)* of the HACCP plan development and implementation: *5 Management responsibility (5.2 Customer focus, 5.4 Planning), 7 Product realization.*

5.6.7 Step 7 — determine the critical control points (HACCP Principle 2)

With the information obtained from the hazard analysis step, the HACCP team must then determine the points at which there will be control of the hazards that present unacceptable risks; this will establish the CCPs of the HACCP plan. A useful tool for the determination of whether a raw material or process step is a CCP, is the CCP decision tree. Examples of CCP decision trees have been proposed by Codex Alimentarius and by the NACMCF. The CCP decision trees consist of a set of either three or four questions, which are asked in a particular sequence for each identified hazard so that the point of control of that hazard within the HACCP plan can be determined. Tables 5.3a and 5.3b summarize the questions that are asked in the decision trees proposed by Codex Alimentarius Commission and the NACMCF, respectively. When the Codex Alimentarius decision tree (Table 3a) is used for determining whether a raw material or ingredient is a CCP, Question 2 does not apply.

Table 5.2 Some Examples of Control Measures for
Biological, Chemical, and Physical Hazards

Control measures for biological hazards:
– thermal processing to eliminate pathogens
– frozen storage to prevent pathogens
– use of preservatives to prevent pathogens
– testing for the presence of pathogens

Control measures for chemical hazards:
– formulation control of regulated food additives
– testing for the presence of antibiotics
– testing for the presence of pesticide residues

Control measures for physical hazards:
– filtering or screening to remove foreign objects
– detection and removal of metal contaminants

Relationship to ISO 9000: The following clauses of *ISO 9001:2000* can be related to *Step 6* of the HACCP plan development and implementation: 7 *Product realization (7.1 Planning of product realization)*

At the completion of *Step 7*, the HACCP team should be in a position to determine at which points in the process, the identified biological, chemical and physical hazards associated with the product will be controlled so that the hazard will be eliminated, prevented, or reduced to an acceptable level. It is possible that more than one step could be a critical control point (CCP) for the same hazard (e.g., pasteurization and refrigerated storage can be CCPs for the same microbiological hazard); on the other hand, a single CCP could control more than one hazard (e.g., a screening step in a process can be a CCP for various physical hazards).

Relationship to ISO 9000: The following categories of the *ISO 9001:2000* can be related to *Step 7* of the HACCP plan development and implementation: *5 Management responsibility (5.4.2 Quality planning), 7 Product realization.*

5.6.8   Step 8 — establish critical limits for each CCP (HACCP Principle 3)

For each of the CCPs that have been determined in *Step 7*, the HACCP team must establish critical limits which will serve as the criteria for accepting or rejecting a raw material or ingredient that is a CCP, or a semi-finished or finished product that is obtained at a process step that is a CCP. A critical limit is commonly a maximum value of a parameter that must not be exceeded or a minimum value of a parameter that must be reached at a CCP. At a CCP, the critical limits must be respected for the hazard to be prevented, eliminated or reduced to an acceptable level, and therefore for the product obtained at the CCP to be acceptable. If the critical limits at a CCP are not respected, the product obtained at the CCP will not be acceptable.

Table 5.3a Codex Alimentarius Critical Control Point Decision Tree

---

Question 1: Do preventive control measures exist?
(a) If the answer is Yes, go to Question 2.
(b) If the answer is No, is control at this step necessary for the safety of the product?
    (i) if the answer is No, this step is not a CCP; proceed to the next identified hazard in the process
    (ii) If the answer is Yes, modify the steps in the process or the product, and return to start of Question 1.

Question 2: Is this step specifically designed to eliminate or reduce the likely occurrence of the hazard to an acceptable level?
(a) If the answer is Yes, this step is a CCP
(b) If the answer is No, go to Question 3.

Question 3: Could contamination with the identified hazard(s) at this step, occur in excess of acceptable level(s) or could these hazards increase to unacceptable level(s)?
(a) If the answer is Yes, go to Question 4
(b) If the answer is No, this step is not a CCP; proceed to the next identified hazard in the process.

Question 4: Will a subsequent step eliminate the identified hazard(s) or reduce the likely occurrence to acceptable level(s)?
(a) If the answer is Yes, this step is not a CCP; proceed to the next identified hazard in the process
(b) If the answer is No, this step is a CCP

Source: Codex Alimentarius (1997); with permission.

Relationship to ISO 9000: The following clauses of *ISO 9001:2000* can be related to *Step 8 (HACCP Principle 3)* of the HACCP plan development and implementation: *5 Management responsibility (5.4.2 Quality planning), 7 Product realization.*

5.6.9 Step 9 — establish monitoring procedures for each CCP (HACCP Principle 4)

For each of the CCPs that have been determined, the HACCP team must establish the monitoring procedures which will be used to monitor or measure the parameters at the CCP to determine whether the critical limits are being respected. Monitoring can also reveal whether a trend toward loss of control at the CCP is developing so that appropriate action can be taken to prevent loss of control before it actually occurs.

It is essential that the monitoring procedures be reliable; if the monitoring procedure involves a measurement, the reliability of the method used should be known. Visual inspection and physical and chemical measurements are frequently used as monitoring procedures. In some situations

Table 5.3b NACMCF Critical Control Point Decision Tree

Question 1: Does this step involve a hazard of sufficient likelihood of occurrence and severity to warrant its control?
(a) If the answer is Yes, go to Question 2
(b) If the answer is No, this step is not a CCP; proceed to the next step of the process.

Question 2: Does a control measure for the hazard exist at this step?
(a) If the answer is Yes, go to Question 3
(b) If the answer is No, is control at this step necessary for the safety of the product?
  (i) If the answer is Yes, modify the step, the process or product, and return to start of Question 2
  (ii) If the answer is No, this step is not a CCP; proceed to the next step of the process.

Question 3: Is control at this step necessary to prevent, eliminate or reduce the risk of the hazard to consumers?
(a) If the answer is Yes, this step is a CCP; proceed to the next step in the process.
(b) If the answer is No, this step is not a CCP; proceed to the next step in the process.

Source: NACMCF (1997); with permission.

microbiological analysis may be required; however, this should only be the case if no other appropriate monitoring procedure is available.

The monitoring procedures must address the following points:

- The reliability of the method; in some situations Official Methods or Reference Methods may be required (e.g.; when some chemical or microbiological methods are used)
- The sampling and sample handling techniques, if applicable
- Whether the method is appropriate for monitoring, in some situations continuous monitoring may be required
- The frequency at which the monitoring is done, if the monitoring is not continuous
- The skill and training of personnel responsible for the monitoring activities
- The equipment used in the monitoring procedure; any equipment used for monitoring must be maintained and calibrated, if required (this is also covered in the HACCP prerequisite programs)

The monitoring procedures should be adequately documented to ensure that these points are addressed. In addition, a record must be generated whenever a monitoring procedure is performed; this record shows that the monitoring was done as required, the person who performed the monitoring, and the actual results of the monitoring. The monitoring record is an essential

Table 5.4 Examples of Monitoring Procedures for Some CCPs

CCPs for biological hazards	Monitoring procedures
Pasteurization	Monitoring of temperature and time
Acidification	Measurement of pH
CCPs for chemical hazards	**Monitoring procedures**
Receiving of raw materials	Examination of certificate of analysis
Labeling	Inspection of labeled products
CCPs for physical hazards	**Monitoring procedures**
Filtering	Inspection of filter
Metal detection	Monitoring of product by metal detector

part of an operating HACCP plan for a product since it is the evidence for whether or not the critical limits were respected. Table 5.4 gives examples of monitoring procedures for some CCPs.

Relationship to ISO 9000: The following clauses of *ISO 9001:2000* can be related to *Step 9 (HACCP Principle 5)* of the HACCP plan development and implementation: *7 Product realization (7.1 Planning for product realization, 7.4.3. Verification of purchased product, 7.5.1 Control of production and service provision, 7.6 Control of monitoring and measuring devices), 8 Measurement, analysis and improvement (8.1 General, 8.2.3 Monitoring and measurement of processes, 8.2.4 Monitoring and measurement of product).*

5.6.10   Step 10 — establish corrective action procedures for each CCP (HACCP Principle 5)

During the development of the HACCP plan, the HACCP team must establish procedures to be followed if and when the monitoring of a CCP reveals that the critical limits are not respected (i.e., a deviation occurs), and therefore there is a loss of control of the hazard at the CCP. A product that is obtained at a process step where the CCPs are not respected is a nonconforming product and is likely to be unsafe if consumed. The procedures that are established to prevent unsafe product from reaching the consumer are the corrective action procedures or deviation procedures. This HACCP principle recognizes that in spite of the fact that a CCP is operating in a process, it is possible that there could be loss of control at the CCP and this loss of control will be detected during the monitoring procedure.

The following points must be addressed in the corrective action procedures:

- The specific, immediate action to be taken when it is discovered that the critical limits at a CCP are not respected, and therefore there is a deviation from the critical limits

- Identification of the cause of the deviation from the critical limits at the CCP
- The actions to be taken to correct the cause of the deviation so that the deviation does not recur
- Determination of the period of time for which the deviation occurred and the quantity of nonconforming product that was prepared during this time
- The actions to be taken with respect to the nonconforming product to ensure that it does not reach the consumer
- The importance of keeping records of all the actions taken to address the deviation and the nonconforming product

For all of the points listed here, the corrective action procedure must identify the personnel who are responsible for the actions to be taken when there is a deviation from the critical limits at a CCP.

In the event that a deviation from the critical limits at a CCP is discovered only after the non-conforming product has been shipped from the establishment to the customer, then the HACCP prerequisite program *Traceability and Recall* (Chapter 4) must be used to address this situation.

Relationship to ISO 9000: The following clauses of *ISO 9001:2000* can be related to *Step 10 (HACCP Principle 5)* of the HACCP plan development and implementation: *8 Measurement, analysis and improvement (8.3 Control of nonconforming product, 8.5.2 Corrective action)*.

5.6.11   Step 11 — establish verification procedures for each CCP and for the entire HACCP plan (HACCP Principle 6)

This HACCP principle requires that the HACCP team develop measures that will evaluate (a) the effectiveness of the HACCP plan that has been developed and (b) the effectiveness of the HACCP system on an ongoing basis after its implementation.

- *Verification of initial HACCP plan:* The verification of the initial HACCP plan involves validation to ensure that the critical limits, the monitoring procedures and the corrective action procedures *(HACCP Principles 3–5, Steps 8–10)* established at each CCP are indeed capable of controlling the hazard that is required to be controlled. This validation must be conducted during the development of the HACCP plan; it is therefore an essential part of the HACCP plan development. The verification activity must also demonstrate that every aspect of the HACCP plan that has been developed must be fully and correctly implemented.

- *Verification of the HACCP plan on a continuous basis:* The HACCP plan that has been developed must include activities that will demonstrate the following:

- The measures (monitoring procedures, corrective action procedures, record-keeping procedures) which have been implemented at each CCP, must continue to be effective and must operate as required by the HACCP plan. The most common form of verification is ongoing review of the required records of monitoring and corrective actions by personnel other than those who were responsible for the monitoring activity or the corrective action. Supervisory personnel should carry out this type of verification. Verification activities could also include periodic sampling and analysis to verify the effectiveness of the CCP to control a hazard. Table 5.5 gives examples of verification activities for some CCPs.
- Every aspect of the entire HACCP plan that has been implemented must continue to function as originally intended; this type of verification is typical of a system audit, and includes an independent review of the entire HACCP plan at a predetermined frequency. In the event this verification activity discovers that there are deficiencies in the HACCP plan, these deficiencies must be corrected immediately.

Relationship to ISO 9000: The following clauses of *ISO 9001:2000* can be related to *Step 11 (HACCP Principle 6)* of the HACCP plan development and implementation: *7 Product realization (7.1 Planning for product realization), 8 Measurement, analysis and improvement (8.1 General, 8.2.2 Internal audit, 8.2.4 Monitoring and measurement of product).*

Table 5.5 Examples of Verification Activities for CCPs

CCPs for Biological Hazards	Verification activities
Pasteurization	Review of pasteurization, records, microbiological testing of product periodically
Acidification	Review of pH measurement records, microbiological testing of product periodically
CCPs for Chemical Hazards	**Verification activities**
Receiving of raw material	Review of certificates of analysis, periodic sampling and testing of raw material
Labeling	Review of labeling inspection records
CCPs for Physical Hazards	**Verification activities**
Filtering	Review of filter inspection records
Metal detection	Review of metal detector records

5.6.12   Step 12 — establish record-keeping and documentation procedures for the HACCP plan and the HACCP system (HACCP Principle 7)

This HACCP principle covers all records and documents that are required for all HACCP plans and for the entire HACCP system. In the development of the HACCP plan, the HACCP team must identify which documents will be required and which records will be kept as part of the monitoring procedures, corrective action procedures, and verification procedures *(HACCP Principles 4–6; Steps 9–11)*. In addition, the HACCP team must maintain the documentation that was compiled during the development of the HACCP plan. This documentation will be required for any verification and review of the HACCP plan. The documents which are expected to be maintained for an HACCP plan include:

- The composition of the HACCP team along with assigned responsibility *(Step 1)*
- The description of the food product *(Step 2)*
- The identification of the intended use of the product *(Step 3)*
- Creation of a process flow diagram *(Step 4)*
- The verified process flow diagram *(Step 5)*
- A summary of the hazard analysis, along with the justification for identification and evaluation *(Step 6, HACCP Principle1)*
- A summary of the CCP determination, along with the justification *(Step 7, HACCP Principle 2)*
- An HACCP plan summary with the following information (Figure 5.2 is an example of a commonly used format for a HACCP plan summary):
  - Steps in the process flow diagram that have been identified as CCPs along with the hazard that is to be controlled
  - The critical limits for each CCP *(Step 8, HACCP Principle 3)*
  - The monitoring procedures for each CCP, including the frequency and personnel responsible *(Step 9, HACCP Principle 4)*
  - The corrective action procedures to be followed in the event of a deviation, including the personnel responsible *(Step 10, HACCP Principle 5)*
  - The verification procedures to be performed for each CCP, as well as for the entire HACCP plan, including the frequency and the personnel responsible *(Step 11, HACCP Principle 6)*
  - The records that will be kept for each CCP in the HACCP plan *(Step 12, HACCP Principle 7)*

- Any other relevant supporting documents that were generated during the HACCP plan development (e.g., data from validation of CCPs)

The records which will be generated when the HACCP plan is used are:

CCP	Hazard(s)	Critical Limit(s)	Monitoring	Corrective Actions	Verification	Records

Figure 5.2 Example of format for a HACCP plan summary.

Table 5.6 Examples of Records to be Kept for CCPs

CCPs for Biological Hazards	Records
Pasteurization	Pasteurization monitoring records
Acidification	pH measurement records

CCPs for Chemical Hazards	Records
Receiving of raw materials	Monitoring records for receipt of certificate of analysis
Labeling	Records of inspection of labeling

CCPs for Physical Hazards	Records
Filtering	Records of inspection of filter
Metal detection	Monitoring records of metal detector

Records for All CCPs
Verification records of monitoring activities
Corrective action records when CCPs are not respected

- Records that result from of monitoring of the CCPs
- Records that result from corrective actions and deviations, whenever they occur
- Records that result from verification of CCPs and of the entire HACCP plan

Table 5.6 gives examples of records that must be kept for some CCPs.
Relationship to ISO 9000: The following clauses of *ISO 9000:2000* can be related to *Step 12* (*HACCP Principle 7*) of the HACCP plan development and

implementation: *4.2 Documentation requirements (4.2.3 Control of documents, 4.2.4 Control of records)*.

5.7   Maintenance of the HACCP system

After an HACCP system has been developed and implemented, it must be maintained effectively on a continuous basis. This means that the monitoring procedures, the corrective action procedures (when required), the verification activities, and the record-keeping at each CCP, and for all the HACCP plans in the HACCP system, must operate continuously, and in exactly the manner as they were initially developed and implemented. Any change in any of these activities should only take place after the HACCP coordinator has been informed and has approved the change. For any significant change to the existing HACCP plan activities, the HACCP team should evaluate the change using the same guidelines and principles *(Steps 1 to 12)* that were used in the development of the HACCP system.

The HACCP team should determine whether the HACCP plan for a product needs to be modified after each of the following:

- The intended use of the product has changed
- There is a change in one or more raw materials, ingredients, or packaging materials used for preparing the product
- There is a change in the process for preparing the product
- There is addition, replacement or modification of equipment used in the process

If any of the above changes affect the existing CCPs in the HACCP plan for a product, the HACCP plan should be reviewed to determine whether modifications are required for the critical limits, monitoring procedures, corrective action procedures, verification procedures, and recording-keeping and documentation procedures.

5.8   References

Pierson, M.D., Corlett, D.A., HACCP, *Principles and Applications*, Chapman & Hall, New York, 1992.

Hazard Analysis and Critical Control Point (HACCP) System and Guidelines for Its Application, Codex Alimentarius Commission, Joint FAO/WHO Food Standards Program, 1997, Rome, www.codexalimentarius.net/.

Hazard Analysis and Critical Control Point Principles and Application Guidelines. August 1997, National Advisory Committee on Microbiological Criteria for Foods, *J. Food Prot.*, 61(9), 1246–1259, 1998.

International Standard, ISO 9001, 3rd ed., 2000–12–15, Quality management systems — Requirements, ISO 2000, Geneva.

Reference Database for Hazard Identification, March 1995, Agriculture and Agri-Food
    Canada, Canadian Food Inspection Agency.
The Quality Auditor's HACCP Handbook, ASQ Food, Drug, and Cosmetic Division,
    ASQ Quality Press, Milwaukee, 2002.

Index

A

Acceptable quality level (AQL), 2
Acceptance sampling, 2
Additives, permitted, 36
Aflatoxins, 36
Agricultural residues
Airborne contamination, 97
Alkaloids, 36
Allergens, 118
American Society for Quality (ASQ), 1, 2
Antibiotics, 37
AQL, *see* Acceptable quality level
Ascaris lumbricoides, 35
ASQ, *see* American Society for Quality
Attributes, method of, 2
Audit(s), 3
    client, 3
    conclusion, 3
    criteria, 4
    evidence, 4
    findings, 4
    food safety, 88
    internal, 57, 81, 82
    program, 4
    team, 4
    types of, 65
Automotive industry, QS 9000 standard of, 44

B

Bacteria
    counts of total, 29
    pathogenic, 34, 35
Benchmarking, 4, 61
Best-in-class companies, 4
Best practice, 4
Biological hazards, 34, 129, 132, 135, 137
Blemish, 4

Body of knowledge (BOK), 4
BOK, *see* Body of knowledge
Breakthrough improvement, 4

C

Calibration, 4
Campylobacter jejeuni, 35
Canada, Food and Drugs Act, 32
Canadian Food Inspection Agency, Food
    Safety Enhancement Program, 30
Cause-and-effect diagram, 5
CCP, *see* Critical control point
Certified Mechanical Engineer, 2
Certified Quality Auditor (CQA), 2
Certified Quality Engineer (CQE), 2
Certified Quality Improvement Associate
    (CQIA), 2
Certified Quality Manager (CQM), 2
Certified Quality Technician (CQT), 2
Certified Reliability Engineer (CRE), 2
Certified Six Sigma Black Belt (CSSBB), 2
Certified Software Quality Engineer (CSQE), 2
CFR, *see* U.S. Code of Federal Regulations
Characteristics, food quality, 30
Check sheet, 5
Chemical hazards, 34, 36, 129, 132, 135, 137
Chemicals, nonfood, 114
CI, *see* Continuous improvement
CIP Equipment, *see* Cleaned-in-Place
    Equipment
Cleaned-in-Place (CIP) Equipment, 105
Cleaned-out-of-Place (COP) Equipment, 105
Cleaning
    chemical compounds used for, 106
    effectiveness, 106
    personnel, 106
    records, 107
    schedule, master, 105